LIFE STORY

BIOLOGY FOR SCHOOL & COLLEGE

TEACHER'S GUIDE

F M SULLIVAN

Oliver & Boyd

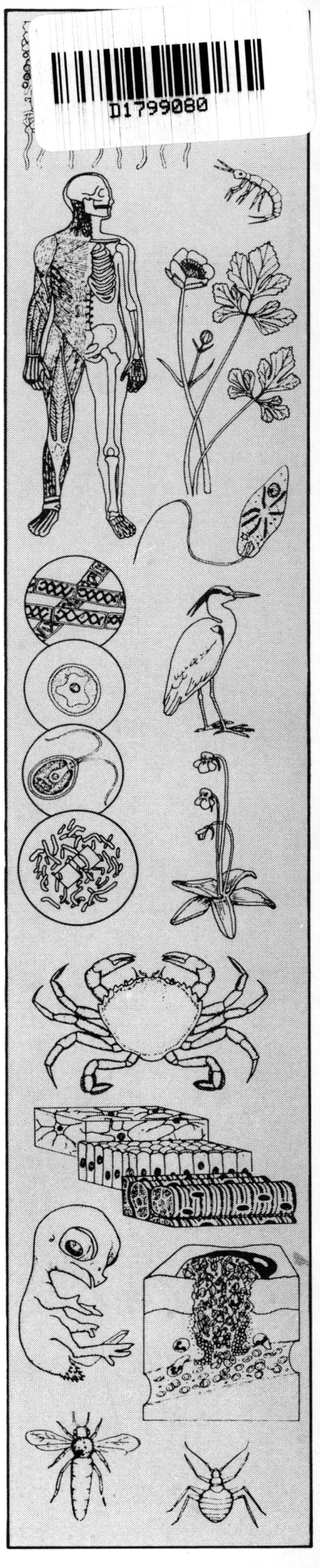

Oliver & Boyd
Robert Stevenson House
1-3 Baxter's Place
Leith Walk
Edinburgh EH1 3BB

A Division of Longman Group UK Ltd

ISBN 0 05 003724 2

First published 1987

© F. McCahill 1986

Printed in Great Britain
by Dotesios Printers Ltd,
Bradford on Avon, Wiltshire.

Contents

Pages

Introduction   5

Preparation of reagents, solutions and materials   6

Experiments 1.1 - 1.3   9-12

Experiments 2.1 - 2.7   12-20

Experiments 3.1 - 3.6   20-29

Experiments 4.1 - 4.10   29-45

Experiments 5.1 - 5.6   45-51

Experiments 6.1 - 6.9   51-62

Experiments 7.1 - 7.11   62-75

Experiments 8.1 - 8.4   75-79

Experiments 9.1 - 9.7   79-85

Experiments 10.1 - 10.11   85-97

Experiments 10.1 - 10.11   85-97

Experiments 11.1 - 11.7   97-104

Experiments 12.1 - 12.8   104-112

Experiments 13.1 - 13.3   112-116

Experiments 14.1 - 14.5   116-120

Introduction

The experiment guide

The purpose of the Experiment Guide is to provide students with clear step by step instructions which will help them to complete each exercise successfully. It is not intended, however, to be a 'Do-It-Yourself' manual. Each experiment should be introduced, outlined and supervised by a teacher. The Experiment Guide will then serve as a reference and reinforcement for the student, eliminating the need for constant repetition of instructions.

Using the experiment guide

Before embarking on an experiment, it is essential that students read through the whole experiment at least once. This will enable them to anticipate their needs for apparatus and materials. They may then collect everything necessary from, and return it to, a central point. Lists of material etc needed to perform each experiment (assuming groups of two) are given in this Guide. It is not advisable for students to read and attempt to follow one instruction at a time without having first obtained a clear overall picture of the whole experiment.

The results

As soon as results are obtained, they should be noted in the Student Record. The teacher can then check them and discuss any anomalies with the student. To facilitate this, a 'Specimen result' is given where appropriate in this Guide.

Think about it!

Once the results have been recorded, the student should attempt to answer the 'Think about it!' sections in a separate jotter. These sections are intended to stimulate necessary revision, to encourage thought about the methods used and to lead on to a suitable conclusion. Answers to the 'Think about it!' sections are given in this Guide.

Conclusions

'Suggested conclusions' are given in this Guide, but the ideal is that the student should be able to produce, in his or her own words, a conclusion which sums up the lessons learned from each experiment.

Preparation of reagents, solutions and materials

Iodine solution

Dissolve 1 g of potassium iodide in 300 ml of distilled water. Add 1 g of resublimed iodine and dissolve.

Benedict's solution

Dissolve 173 g of sodium citrate and 90 g of anhydrous sodium carbonate in 850 ml of distilled water and filter. This is solution A. Dissolve 17.3 g of copper sulphate in 150 ml distilled water. This is solution B. Slowly add solution B to solution A, stirring constantly. This solution will keep if stored. (Keep away from skin. Do not swallow.)

Fehling's solution

Dissolve 35 g of copper sulphate in 1000 ml distilled water. This is solution A. Dissolve 173 g of sodium potassium tartrate and 120 g of sodium hydroxide in 1000 ml distilled water. This is solution B. Keep solutions A and B separate until needed, then mix in equal volumes.

1% starch solution (1)

Measure out 100 ml of distilled water. Use a little of this to make a paste with 1 g starch. Bring the rest to the boil, then add the paste, stirring constantly. Boil for one minute, then allow to cool.

1% starch solution (2)

Measure out 250 ml of distilled water. Use a little of this to make a paste with 2.5 g of starch and 0.25 g of sodium chloride. Bring the rest to the boil, then add the paste, stirring constantly. Boil for one minute, then allow to cool.

2% starch solution

Measure out 250 ml of distilled water. Use a little of this to make a paste with 5 g of starch and 0.25 g of sodium chloride. Bring the rest to the boil, then add the paste, stirring constantly. Boil for one minute, then allow to cool.

Starch-glucose solution

Measure out 100 ml of distilled water. Use a little of this to make a paste with 5 g of starch. Bring the rest to the boil, then add the paste, stirring constantly. Add 10 g glucose and stir until dissolved. Allow to cool.

Egg albumen suspension

Pour the white of an egg into its own volume of distilled water. Beat the mixture until frothy, then slowly pour it into 500 ml of distilled water at 60°C, stirring constantly. Gently heat the mixture until it becomes opaque. Do not heat it above 80°C.

Bicarbonate indicator solution

Dissolve 0.84 g of sodium bicarbonate in about 800 ml of distilled water. Dissolve 0.1 g of cresol red and 0.2 g of thymol blue in 20 ml of ethanol. Add the dye solution to the bicarbonate solution and make up to exactly 1000 ml. Keep this as the stock solution.

Before use, make up 25 ml of the stock solution to exactly 250 ml with distilled water. This solution should be deep red. If it is not, use an aquarium pump, fitted with a plastic delivery tube and a diffuser, to pump fresh air through the solution until it goes deep red.

Lime water

An excess of solid calcium hydroxide is kept at the bottom of a Winchester full of water. The day before the lime water is required, top up the calcium hydroxide with water and shake it. The lime water can be siphoned off when required without disturbing the sediment.

Paramecium culture medium

Boil 10 g of chopped straw or hay in 1000 ml of distilled water for 15 minutes. Cool, then add three boiled and crushed wheat grains. Leave for a day, then filter into a clean flask. Adjust the pH to between 6 and 7, if necessary, using dilute NaOH. Shake vigorously to aerate, then inoculate with culture containing Paramecium caudatum. Keep in a warm place in dim light. Subculture, in volumes of about 250 ml, every 3 or 4 weeks. Before use, centrifuge to concentrate the organisms. (All utensils and vessels should be kept scrupulously clean and used only for handling the cultures.)

Nutrient agar plates

Dissolve 5 g of agar and half an Oxo cube (or a small teaspoonful of Bovril) in 250 ml distilled water. Leave for 15 minutes. Heat very slowly, stirring constantly, until the liquid just starts to boil. Stir until the agar is dissolved, pour into a flask, plug with cotton-wool, then autoclave for 20 minutes. When preparing plates, heat the flask in a water bath until the agar melts, then pour it into sterile Petri dishes until the base is just covered - the lid should be raised as little as possible while pouring. Allow to cool. Dry the agar by stacking the plates as shown below and incubating at 37°C for 15-30 minutes. Keep the agar inverted throughout.

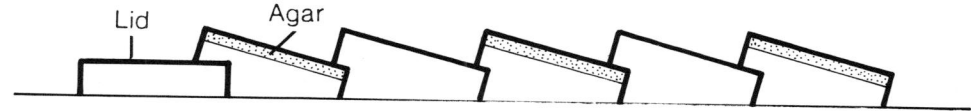

Replace each lid as the plates are lifted.

Nutrient agar and prepared plates can also be purchased from Oxoid, Difco and other biological suppliers.

Nutrient broth cultures of bacteria

Obtain nutrient broth tablets from a supplier. Put each tablet in 10 ml of distilled water in a McCartney bottle. Screw on the caps loosely and autoclave for 15 minutes. Allow to cool. Using sterile technique, inoculate each bottle with bacteria recently bought from a biological supplier. These may be in the form of impregnated paper discs or on agar slopes. Label each bottle with the name of the bacteria and the date of inoculation, then incubate. After use, autoclave and dispose of the broth.

Fungal cultures on agar slopes

Prepare nutrient agar. Put about 10 ml into each McCartney bottle. Screw on the caps loosely and autoclave for 20 minutes. Tighten the caps and prop the bottles at an angle on the bench until the agar sets into a slope. Using sterile technique, draw a loopful of a purchased fungal culture across the surface of the agar in each bottle. Label each bottle with the name of the fungus and the date of inoculation, then incubate. After use, autoclave and dispose of the agar.

Indole acetic acid

Measure out 5 ml of ethanol and dissolve 0.2 g of crystalline IAA in it. Put 20 g lanolin in a small, wide-mouthed bottle and melt it by placing the bottle in warm water. Add the IAA/ethanol and stir it for a few minutes. Label the bottle (include the date). This can be kept for a year or longer in a refrigerator.

1.1   Investigating cells

Object

To examine a variety of plant and animal cells under the microscope, estimate their size in microns (micrometres) and draw them

Material per class

Onion separated into layers and cut into 1 cm squares
Plastic spatulas in a beaker of disinfectant, eg 1% solution of an ampholytic
  disinfectant (rinse before use)
Elodea leaves in water
'Pond water' - perhaps from the bottom of an aquarium, or hay infusion (see page 7)
Polycell solution, equal parts Polycell and water, with a dropper
One-week-old mustard or cress seedlings (Keep moist.)
Small pieces of fresh meat
Graticules on microscope slides (These can be bought ready-made or prepared from
  packs supplied by Philip Harris. A piece of Scalafix tape stuck on a slide could
  also be used.)
Prepared slides of blood, root, stem, leaf, etc
Flowers

Material per group/pupil

Microscope, with bench lamp if necessary   Mounted needle
Microscope slides   Iodine solution in a labelled dropping
Cover slips   bottle (see page 6)

Comments

It should be emphasised that when the slide and objective are being brought closer together, the pupils must look from the side to make sure they never touch.

Specimen result

None

Suggested conclusion

Cells come in a variety of shapes and sizes, though they all have certain features in common.

1.2   Mitosis in a root tip

Object

To prepare and stain a root tip squash so that cells in various stages of mitosis may be seen under the microscope

Material per class

Root tips of onion, mustard, hyacinth, broad bean - A firm, clean onion in a warm place will provide suitable root tips in 1-2 days. Several should be set up in case of failure. A few scalpel cuts in the base of the onion will help growth. For better results, the root tips can be fixed by placing them for 5 minutes in a mixture of:
  10 volumes of absolute alcohol;
   2 volumes of glacial acetic acid;
   2 volumes of chloroform;
   1 volume of formalin.
They can then be put in water and given out.
Propionic orcein or acetic orcein stain in a labelled dropping bottle (Caution - this causes burns.)
1M hydrochloric acid in a labelled dropping bottle.
Prepared root tip squash slides if necessary

Material per group/pupil

Test tube   Cover slip
Beaker, thermometer and hot water from   Filter paper strips
  tap or a thermostatic water bath set   Mounted needle
  at 60°C   Microscope, with bench lamp if necessary
Microscope slide

Specimen result

None

Think about it!

1   The cells are not all the same size.
2   The smallest cells are just behind the tip.
3   Small cells should be younger than large cells.
4   New cells are being formed just behind the tip.
5   They are newly formed by cell division and so have only half the usual cytoplasm.
6   Chromosomes are visible in certain cells.
7   These cells are just behind the tip.
8   I can see ... kinds of different-looking nuclei.
9   These cells are in the following stages of mitosis:
      (i), etc.
10   The region of cell division in a root is just behind the tip.

Suggested conclusion

Mitosis and cell division in a root take place just behind the tip. Only in that region can chromosomes be seen. I saw cells at the following stages of mitosis there: (i) ....., etc.

1.3 The growing region in a root

Object

To find the region of elongation in a root

Material per class

Soaked broad beans, at least 2 per group
Paper towels or newspaper

Thread or elastic bands
Tall beakers or jars of water

Material per group/pupil

Soft tissues
Watch glass with Indian ink
Blotting paper
Hair grip or paper clip with thread
    stretched across it
Large Petri dish

Cotton-wool strips about 30 mm wide,
    the full thickness of the roll
Elastic band
Dropper or dropping bottle of water
Label or marker

Comments

The beans will take 6 to 10 days to produce roots of a suitable length. After marking, a result should be obtained within 3 or 4 days.
Peas, maize, runner beans are also suitable.
Most elongation occurs within 2 or 3 mm of the tip.

Specimen result

None

Think about it!

1   If the roots grow, the distance between the marks will widen.
2   This happened to the marks just behind the tip.
3   Growth in length takes place just behind the tip of a root.
4   The vacuole takes up most of a mature plant cell.
5   Root growth could also come about by cells developing vacuoles.

Suggested conclusion

The region of growth in a root is just behind the tip. This is called the region
of elongation. Cells there are developing vacuoles and stretching.

2.1   Plants, light and starch(1)

Object

To show that starch is present in green leaves only when they have been exposed to
light

Material per class

Two healthy geranium plants, one having been well illuminated and the other kept
   in total darkness for 2 days. (To save time and avoid too much damage to the
   plants, the teacher can cut the pieces of leaf and make them available to the
   class.)
Large labelled beaker for used alcohol. The alcohol and leaves are dumped together
   into this and the leaves then retrieved with forceps; the alcohol can be redistilled
   for future use.

Note: the clear industrial methylated spirits is preferable to the coloured
mineralised methylated spirits, though, in any case, good ventilation is essential.

Material per group/pupil

Beaker   Petri dish lid or watch glass
Test tube   Dilute iodine solution in a labelled
Bunsen   dropping bottle (see page 6)
Tripod/wire gauze   Forceps

Note: to reduce fire hazard a thermostatic water bath set at 80°C may be used
instead of a bunsen.

Specimen result

Plant	Light or dark?	Had it made starch?
A ▲	Light	Yes
B ■	Dark	No

Suggested conclusion

Plants need light to make starch (a kind of food).

2.2   Plants, light and starch(2)

Object

To show that starch is formed only in those parts of a leaf which have been exposed to light

Material per class

Healthy pot plants, preferably one per group. These are destarched by being deprived of light for 2 days. Geranium, french bean and busy Lizzie (<u>Impatiens</u>) are all suitable.

Fluorescent tubes set up to provide close illumination - about 5 hours should provide a result.

Large labelled beaker for used alcohol. The alcohol and leaves are dumped together into this and the leaves then retrieved with forceps. It is essential that the leaf be completely decolourised before iodine is added. If necessary, top up the tube with fresh alcohol. The used alcohol can be redistilled for future use.

<u>Note</u>: the clear industrial methylated spirits is preferable to the coloured mineralised methylated spirits, though, in any case, good ventilation is essential.

Material per group/pupil

Black paper or aluminium foil
Paper clips
Beaker
Test tube
Bunsen
Tripod/wire gauze

Petri dish lid or watch glass
Dilute iodine solution in a labelled
    dropping bottle (see page 6)
Methylated spirits
Forceps

<u>Note</u>: to reduce fire hazard a thermostatic water bath set at $80^{o}C$ may be used instead of a bunsen.

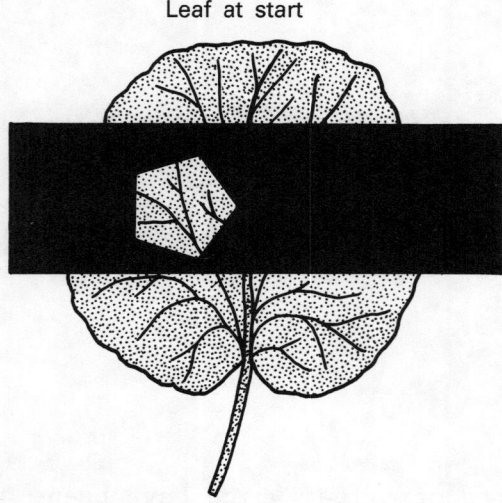

Leaf at start

Leaf after testing

Think about it!

1   The parts of the leaf not covered with paper contained starch.
2   The black paper kept out the light.
3   Maybe the fact that the leaf was being touched had the effect. Perhaps try with clear film.

Suggested conclusion

Plants need light to make starch (a kind of food).

2.3   Plants, sugar and starch

Object

To show that leaves can convert glucose to starch

Material per class

A healthy geranium plant. This is destarched by being deprived of light for 2 days. (To save time and avoid too much damage to the plant the teacher can cut the pieces of leaf and make them available to the class.)

Large labelled beaker for used alcohol. The alcohol and leaves are dumped together into this and the leaves then retrieved with forceps. The alcohol can be redistilled for future use.

Note: the clear industrial methylated spirits is preferable to the coloured mineralised methylated spirits, though, in any case, good ventilation is essential.

Material per group/pupil

2 test tubes with stoppers or Parafilm
Labels
Beaker
Tripod/wire gauze
Forceps
5% glucose solution

Methylated spirits
Petri dish or watch glass
Dilute iodine solution in
    a labelled dropping bottle
    (see page 6)

Note: to reduce fire hazard a thermostatic water bath set at $80^{\circ}C$ may be used instead of a bunsen.

Specimen result

Leaf	Liquid bathed in	Did it contain starch?
A ■	Water	No
B ▲	5% glucose Solution	Yes

Think about it!

1   A starch molecule is a chain of hundreds of glucose molecules joined together.
2   Starch was made in this experiment.
3   The starch could not have been made by photosynthesis since there was no light.
4   The starch could have been made from the glucose.
5   Leaf A shows that without glucose no starch is made.
6   Sugar made by photosynthesis in a leaf could be converted immediately to starch.

Suggested conclusion

Leaves can convert glucose (sugar) to starch. So, if they make sugar by photosynthesis, they can quickly change it to starch.

2.4   Plants, carbon dioxide and starch

Object

To show that starch is produced in illuminated green leaves only when carbon dioxide is available

Material per class

Two healthy geranium plants. These are destarched by being deprived of light for 2
days. They are then illuminated for up to 2 days (though as little as 5 hours may
be enough) with plant A deprived of carbon dioxide and plant B supplied with it.
Concentrated potassium hydroxide may be used instead of soda-lime, but it is more
awkward to handle. (To save time and avoid too much damage to the plants the
teacher can cut the pieces of leaf and make them available to the class.)

Note: the result can be guaranteed if plant A is secretly kept in darkness throughout.
This is unethical.

Large labelled beaker for used alcohol. The alcohol and leaves are dumped together
into this and the leaves then retrieved with forceps. The alcohol can be
redistilled for future use.

Note: the clear industrial methylated spirits is preferable to the coloured
mineralised methylated spirits, though, in any case, good ventilation is essential.

Material per group/pupil

Beaker   Petri dish lid or watch glass
Test tube   Dilute iodine solution in a
Bunsen   labelled dropping bottle
Tripod/wire gauze   (see page 6)
Methylated spirits   Forceps

Note: to reduce fire hazard a thermostatic water bath set at $80^{o}C$ may be used
instead of a bunsen.

Specimen result

Plant	CO_2/No CO_2?	Had it made starch?
A ■	No CO_2	No
B ▲	CO_2	Yes

Suggested conclusion

Plants need carbon dioxide to make starch (a kind of food).

2.5 Plants, chlorophyll and starch

Object

To show that starch can be detected only in those parts of a leaf which contain
chlorophyll

16

Material per class

A healthy, well-illuminated, variegated plant. If a wandering sailor (<u>Tradescantia fluminensis</u>) is used, it may be possible to give a leaf to each pupil. Then, after testing, the leaf can be washed, dried with blotting paper and stuck, under clear adhesive tape, beside the original drawing in the Student Record.

Large labelled beaker for used alcohol. The alcohol and leaf are dumped together into this and the leaf then retrieved with forceps. It is essential that the leaf be completely decolourised before iodine is added. If necessary, top up the tube with fresh alcohol.

<u>Note</u>: the clear industrial methylated spirits is preferable to the coloured mineralised methylated spirits, though, in any case, good ventilation is essential.

Material per group/pupil

Beaker

2 test tubes

Bunsen

Tripod/wire gauze

Methylated spirits

Petri dish lid or watch glass

Dilute iodine solution in a labelled
  dropping bottle (see page 6)

Forceps

<u>Note</u>: to reduce fire hazard a thermostatic water bath set at 80°C may be used instead of a bunsen.

Specimen result

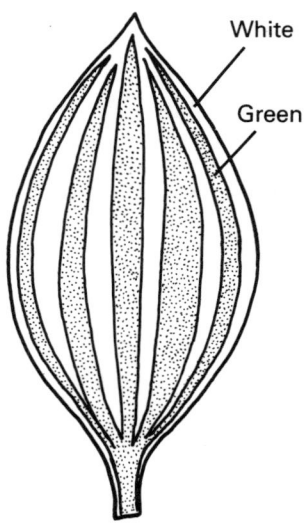

White

Green

Leaf before testing

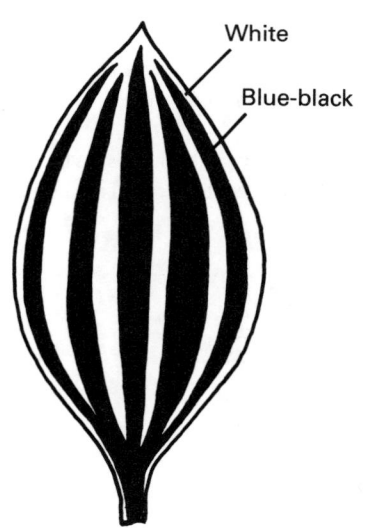

White

Blue-black

Leaf after testing

Suggested conclusion

Plants need the green substance chlorophyll to make starch (a kind of food).

2.6 Chloroplasts

Object

<u>Method A</u> To show that chlorophyll is located inside the chloroplasts of a leaf and that these contain starch grains produced in photosynthesis

<u>Method B</u> To show that the green colour and hence the chloroplasts are concentrated at the upper surface of a dicotyledon leaf

Material per group/pupil

<u>Elodea</u> or moss leaves
Microscope and lamp
Microscope slides
Cover slips
Forceps
Mounted needle
Dropper or dropping bottle of water

Privet leaves
Carrot or pith
Scalpel or razor
Iodine solution in labelled
  dropping bottle (see page 6)
Prepared slide of TS of leaf if
  necessary

Comments

<u>Method A</u> Because the leaves are more than one cell thick, background colour can be confusing. To convince the pupil that the colour is, in fact located inside the chloroplasts, cells near the edge, if not too full of chloroplasts, should be best.

<u>Method B</u> Cutting thin sections is difficult and it may be impossible to make out much detail of cells. In this case, a simple tissue map is adequate.

Specimen result

See <u>Experiment Guide</u>, Section 2.

Think about it!

1   Chlorophyll is found inside the chloroplasts in a plant cell.
2   Starch grains are found inside the chloroplasts in a plant cell.
3   Chloroplasts contain the chlorophyll needed for photosynthesis and the starch produced by it.
4   The tissue, called the palisade layer, near the top of a dicotyledon leaf contains most chloroplasts.
5   Being at the top of the leaf, the chloroplasts catch lots of light for photosynthesis.

Suggested conclusion

Chloroplasts are the sites of photosynthesis in a leaf. They contain the chlorophyll needed for it and the starch made by it. The palisade layer in a leaf, which contains most chloroplasts, is at the top to catch the light.

2.7   The waste gas of photosynthesis

Object

To show that plants give off oxygen in the light

Material per group/pupil

Healthy sprig of <u>Elodea</u> or other aquatic   Wood splint
   plant   Stopper
Glass filter funnel   Label
Large beaker or trough of water
Test tube

Comments

Adding a pinch of potassium bicarbonate will help provide the plant with carbon
dioxide. Some people help a sluggish plant by secretly topping up with oxygen
from a cylinder. This is unethical.

Specimen result

An invisible gas collected in the test tube. This gas relit a glowing wood splint.

Suggested conclusion

Plants give off oxygen in the light.

2.8   Light intensity and photosynthesis

Object

To show that the rate of photosynthesis is directly proportional to light intensity
at lower light levels

Material per class

Light meters
1% potassium bicarbonate solution

Fresh, healthy shoots of <u>Elodea</u> with ends freshly cut, in a tank. They should be
  well illuminated and as many as possible should be prepared so that the best
  bubblers can be selected.

Material per group/pupil

Large boiling tube   Stopclock
Stand fitted with clamp   Bench light(s)
Large beaker of water

Comments

Thin tissue paper may also be used to screen the plant to help vary light intensity.
The temperature should be checked occasionally to ensure it is constant.
The size of bubbles can vary a lot. A small piece of drawn-out glass tubing fitted
loosely to the end of the shoot helps to regulate this.

Specimen result

Light intensity units	Bubbles of oxygen per minute
0	0
60	6
120	12
220	21
320	27
390	30

Think about it!

1   Increasing the light intensity increases the rate of bubbling.
2   Increasing the light intensity increases the rate of photosynthesis since the
     bubbles given off are oxygen which is produced in photosynthesis.
3   The plant needs carbon dioxide for photosynthesis.
4   Temperature affects the rate of photosynthesis.

Suggested conclusion

Increasing the light intensity increases the rate of photosynthesis (up to a point).

3.1   Enzymes:biological catalysts

Object

To show that potato tissue contains an enzyme (catalase) which can catalyse the
breakdown of hydrogen peroxide

Material per class

Hydrogen peroxide (20 volumes). This is supplied with an inhibitor to hinder
  decomposition. Before use, this can be precipitated out with a few drops of
  sodium hydroxide which is then filtered off.
Clean sand
Manganese dioxide
10 ml syringes

Material per group/pupil

3 test tubes   3 potato cylinders, about 3 cm long,
Test tube rack   cut with a cork borer
Wood splint

Comments

A few drops of washing up liquid will stabilise the foam.
If a glowing wood splint is thrust quickly into the foam, it will relight before
being extinguished.

Specimen result

Test tube	Treatment	Did it froth?	Glowing splint?
A	Sand	No	Nothing
B	Manganese dioxide	Yes	Relit
C	Potato	Yes	Relit

Think about it!

1   A catalyst would make it break down faster.
2   Manganese dioxide speeded it up. It is a catalyst for this reaction.
3   The manganese dioxide did not seem to be used up. We would not expect a
     catalyst to be used up.
4   No, sand is a powder and it had no effect.
5   Biological catalysts are called enzymes.

Suggested conclusion

Potato contains an enzyme which speeds up the breakdown of hydrogen peroxide into
oxygen and water. This enzyme is called catalase.

3.2 Enzymes and living tissues

Object

To show that enzymes can be found in a wide variety of plant and animal tissues.

Material per class

Hydrogen peroxide solution. Dissolve 20 volume hydrogen peroxide solution with 4 times its own volume of water.

Note: the term '20 volume' means that this solution can produce 20 times its own volume of oxygen.

A variety of fresh plant and animal tissues
Scalpels or knives and bench mats for cutting on

Material per group/pupil

Test tubes - one for each tissue
Test tube rack

Specimen result

Tissue tested	Did it froth?	Amount of catalase
Blood	Yes	10
Liver	Yes	9

Think about it!

1   It will relight a glowing wood splint.
2   This enzyme seems to be very widespread.
3   Different tissues contain different amounts of catalase.

Suggested conclusion

Catalase is an enzyme present in many different plant and animal tissues, though some contain more than others.

3.3   Phosphorylase: a 'building-up' enzyme

Object

To obtain a sample of phosphorylase from potato and show that it can synthesise starch from glucose-1-phosphate

Material per class

Clean sand
Centrifuge(s)
1% glucose-1-phosphate solution in a labelled dropping bottle. This should be made
   up in minimum quantity within one hour of use.
Fresh potato
Scalpels or knives with bench mats for cutting on.
Distilled water and dropper.

Material per group/pupil

Mortar and pestle
2 centrifuge tubes
Test tube rack
Dropper

Spotting tile
Iodine solution in a labelled dropping
   bottle (see page 6)
Stopclock

Comments

The potato should be ground with the minimum of water.
Pouring into the centrifuge tubes should be done very slowly to hold back as much solid material as possible.
When removing extract from the centrifuge tube, great care must be taken not to disturb the sediment on the bottom.

Specimen result

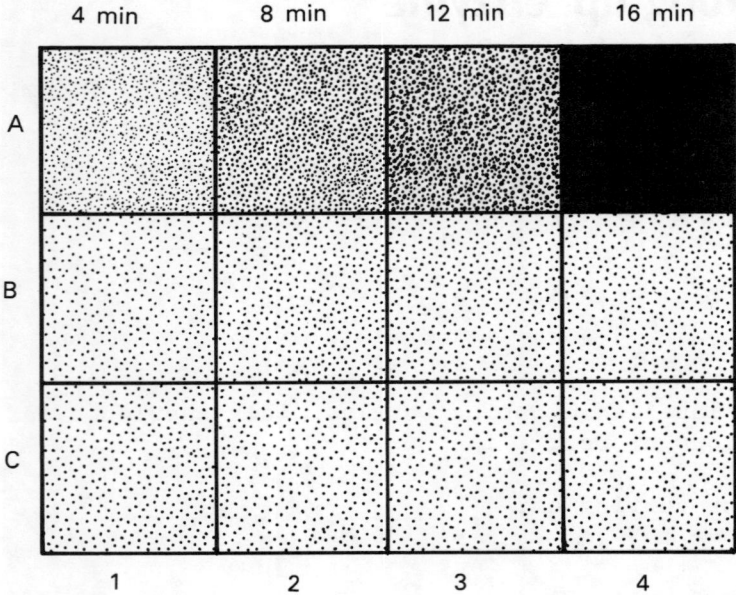

Think about it!

1 Starch turns iodine blue-black.
2 Potato tissue does contain starch.
3 The starch was removed by centrifuging.
4 Starch was formed in row A.
5 A starch molecule is a chain of hundreds of glucose molecules joined together.
6 Glucose-1-phosphate does not turn into starch by itself. Row C contained glucose-1-phosphate but no starch formed.
7 Potato extract does not produce starch by itself. Row B contained potato extract but no starch formed.
8 Potato extract contains an enzyme called phosphorylase which can convert glucose-1-phosphate into starch.

Suggested conclusion

Potatoes contain an enzyme which can convert the sugar, glucose-1-phosphate, into starch. This enzyme is called phosphorylase.

3.4 Amylase: a 'breaking-down' enzyme

Object

To obtain a sample of amylase and show that it can digest starch to form reducing sugar

Material per class

5-day-old barley seedlings, enough for a small handful for each group. The seeds
   are soaked for a few hours to soften them, then placed in seed trays on wet paper.
   They should be turned as often as possible to prevent disease.
Muslin
1% starch solution (see page 6) decanted into small beakers
2 ml syringes.

Material per group/pupil

Mortar and pestle
3 test tubes
Test tube rack
Labels
100 ml beaker
5 ml syringe
Spotting tile
3 droppers

Iodine in a labelled dropping bottle
   (see page 6)
Benedict's solution in a labelled
   dropping bottle (see page 6)
Large beaker   } or a thermostatic
Tripod/wire gauze } water bath
Bunsen   } at 90°C
Stopclock

Comments

It may be more convenient for the teacher to make the extract up beforehand.
Alternatively, 1% diastase or amylase solutions could be prepared from enzymes
obtained commercially.
As an alternative to a numerical result, the Result table in the Student Record
may be coloured in to show the appearance of the spotting tile.

Specimen result

Test tube	Amount of starch (0-5)						Sugar (absent/present)
	0 min	5 min	10 min	15 min	20 min	25 min	
A	5	4	3	2	1	0	Present
B	5	5	5	5	5	5	Absent
C	0	0	0	0	0	0	Absent

Think about it!

1   Test tube A contained starch - it turned iodine blue-black.
2   Test tube A contained no starch at the finish - the iodine did not change
     colour.
3   Test tube A contained sugar at the finish - when heated with Benedict's
     solution, it turned orange.
4   A starch molecule is a chain of hundreds of glucose molecules joined together.
5   The sugar could have come from the starch molecules being broken up.
6   Something in the extract caused the starch to disappear because without it
     (test tube B) nothing happened.
7   The extract contained no sugar since test tube C contained none.
8   The extract could not make sugar without starch since test tube C contained none.

Suggested conclusion

The extract contained an enzyme which broke down starch molecules into the sugars of which they are made. This enzyme is called amylase.

3.5   The effect of pH on the activity of catalase

Object

To show that pH affects the activity of catalase in potato tissue and to estimate its optimum pH

Material per class

5 bottles of 20 volume hydrogen peroxide solution adjusted to different pHs. The
  solutions should be made of fresh stock as soon as possible before use. The pH
  is adjusted by adding dilute hydrochloric acid and dilute sodium hydroxide dropwise,
  checking the pH with a pH meter or pH paper. Suggested values are 1,3,7,9 and 11.
5 labelled beakers for dispensing the solutions
5 5 ml syringes
Fresh potato (It is less wasteful and more convenient for the teacher to cut the
  potato cylinders. Cylinders 8 mm in diameter and about 30 mm long are suitable.)

Material per group/pupil

5 test tubes
5 labels
Test tube rack

Comments

All the test tubes used should be of the same dimensions.
A few drops of washing up liquid will help stabilise the froth.
If there is not much froth, measuring from the bottom of the tube will give a very
flat graph. Measure the actual thickness of the foam instead.
This experiment is not intended to be a rigorous investigation of the optimum pH
of catalase. Rather, it is meant to be a simple, semi-quantitative demonstration
of the idea of optimum pH in enzymes.

Specimen result

pH	Weight of oxygen foam
1	49
3	52
7	65
9	54
11	34

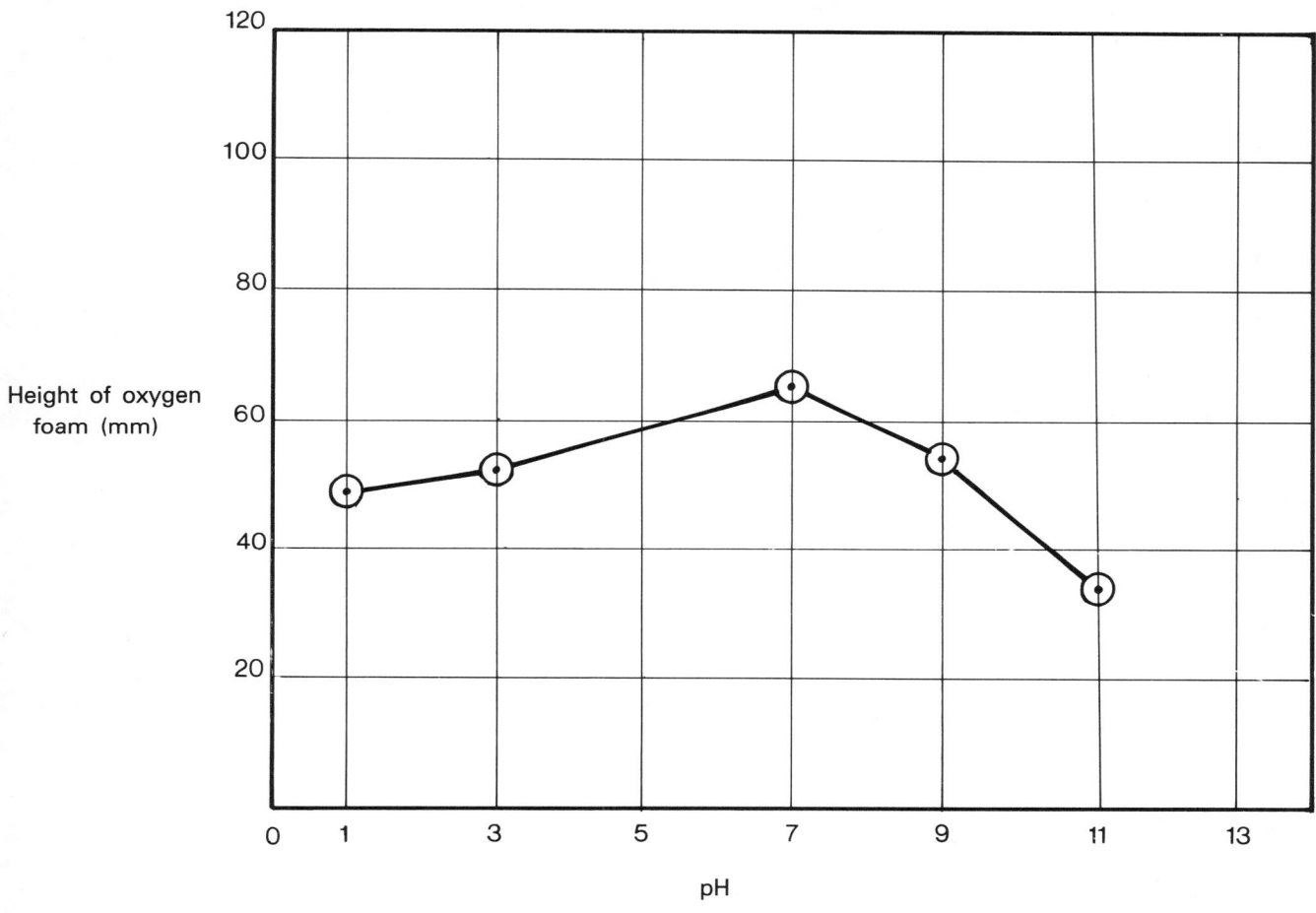

Think about it!

1   Solutions of pH below 7 are acid and above 7 are alkali.
2   Catalase worked best at pH7
3   The best pH of an enzyme is called its optimum pH.
4   Catalase works over a range of pHs.
5   Catalase works over the range pH ... to pH ...
6   The potato cylinders should be the same size to supply the same amount of enzyme.
7   The foam should relight a glowing wood splint.

Suggested conclusion

Catalase works over a range of pH from pH ... to pH ... but seems to work best at pH ... (its optimum pH).

3.6 The effect of temperature on the activity of amylase

Object

To show that rise in temperature increases the activity of salivary amylase up to about $37^{\circ}C$ but thereafter quickly decreases it

Material per class

Water baths set at $0^{\circ}C$, $20^{\circ}C$, $37^{\circ}C$, $55^{\circ}C$, containing test tube racks if necessary (See Experiment Guide, Section 3.)
1% starch solution, 8 ml per group (see page 6)
10% iodine solution with dropper (see page 6)
2 ml syringes

Material per group/pupil

100 ml beaker
4 test tubes
Stopclock

Comments

The results from this experiment can often be patchy, especially if the saliva is too dilute. Undiluted saliva strained through cheesecloth may help. An alternative is to omit the iodine from the starch tube, then test samples with iodine on a spotting tile at regular intervals as in Experiment 3.3.

Specimen result

Temperature	Time taken for colour to go	mg starch digested per minute
$0^{\circ}C$	-	0
$18^{\circ}C$	17 min	1.2
$37^{\circ}C$	10 min	2
$55^{\circ}C$	25 min	0.8

28

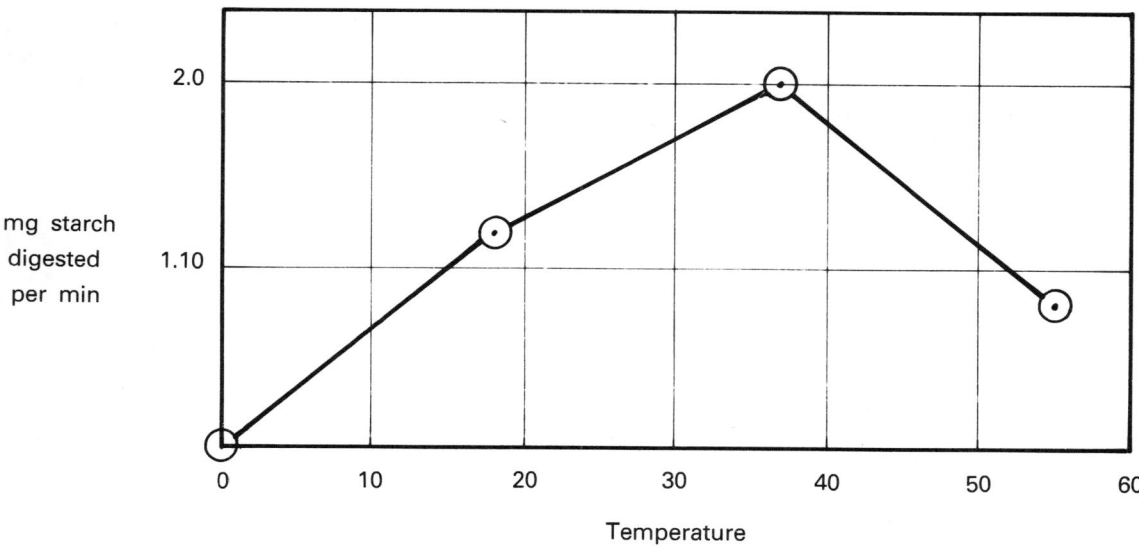

Temperature

Think about it!

1   Amylase works best at $37^{o}C$.
2   This is called its optimum temperature.
3   Amylase worked over a range of temperatures.
4   If an enzyme is heated much above its best temperature, it is denatured and stops working.
5   Yes, after $37^{o}C$ it did not work so well.
6   No, you cannot say for sure that amylase caused the change.
7   Control tubes could be set up containing water instead of saliva.
8   The product of amylase is maltose (a sugar).

Suggested conclusion

Increase in temperature increases the activity of amylase up to its optimum of $37^{o}C$. Above this, activity falls off quickly as the enzyme is denatured.

4.1   Releasing energy from food/Burning foods in oxygen

Object

To show that food is rich in energy and that, when burned in oxygen, this energy is released with the food being converted to carbon dioxide and water.

29

Method A: Teacher demonstration

Comments

The carbohydrate used could be custard or lycopodium powder. If lycopodium is used, it is best described as being similar to custard.
It is better not to blow into the tube by mouth. This may be harmful.
If possible, this demonstration should be given in the dark.

Specimen result

There was a loud bang and a flame and the lid was blown off.

Think about it!

1   The food contained a lot of energy.
2   To start with, the food contained chemical energy.
3   This energy was released as sound, heat, light and kinetic energy.
4   The carbohydrate was very rich in energy.
5   If the cells released energy from food in one step, it would probably kill them.

Suggested conclusion

Food contains a lot of energy. If cells released this suddenly, it would probably kill them.

Method B: Pupil activity

Material per class

Oxygen cylinder with delivery tube
Jars of dry oxygen replenished from the cylinder on the
   return of a clean, dry gas jar and lid
Labelled dishes of dried foods. There should be a variety
   of foods, dried in an oven at $80^{\circ}C$-$90^{\circ}C$ and powdered.

Material per group/pupil

Bunsen burner   Deflagrating spoon
Bench mat   Bottle of fresh lime water (see page 7)

Specimen result

Food burned	Did water form?	Lime water milky/not milky?
Bread	Yes	Milky
Potato	Yes	Milky
Meat	Yes	Milky
Beans	Yes	Milky
Cheese	Yes	Milky
Ham	Yes	Milky

Think about it!

1   There was no oxygen in the jar at the end. We know this because the food stopped burning.
2   Energy was released from the food - heat, light, sound and kinetic energy.
3   Carbon dioxide gas was left at the end.
4   Water was also produced by the reaction.
5   A simple word equation to sum up this result would be:

$$FOOD + OXYGEN \longrightarrow ENERGY + CARBON\ DIOXIDE + WATER$$

Suggested conclusion

When foods are burned in oxygen, their chemical energy is released as heat, light, sound and kinetic energy and the food is converted to carbon dioxide gas and water. It can be summed up like this:

$$FOOD + OXYGEN \longrightarrow ENERGY + CARBON\ DIOXIDE + WATER$$

4.2   The signs of respiration

Object

To show that breathed air contains less oxygen, more carbon dioxide, more water vapour and is warmer than unbreathed air

Material per class

Troughs of water
Drinking straws

Method A: Oxygen content

Material per group/pupil

Gas jar with cover   Stopclock
Long wood splint   Bunsen

Specimen result

	Time of burning
Unbreathed air	8 seconds
Breathed air	1 second

Think about it!

1   Oxygen allows things to burn in it.
2   When oxygen is used up, the burning splint goes out.

Suggested conclusion

Breathed air contains less oxygen than unbreathed air.

Method B: Carbon dioxide content

Material per group/pupil

2 boiling tubes with stoppers
Test tube rack
Air-equilibrated bicarbonate indicator (see page 7)

Specimen result

	Bicarbonate indicator	
	Colour at start	Colour at finish
Unbreathed air	Red	Red
Breathed air	Red	Yellow

Think about it!

1   Carbon dioxide turns bicarbonate indicator yellow.
2   Breathed air has the greater effect on bicarbonate indicator.
3   Breathed air contains more carbon dioxide.

Suggested conclusion

Breathed air contains more carbon dioxide than unbreathed air.

Method C: Water content

Material per group/pupil

Large beaker of crushed ice in water
2 boiling tubes fitted with stoppers,
   glass and plastic tubing as shown in the Experiment Guide. The tubing
   should be cleaned with disinfectant, eg 1% ampholytic disinfectant, and rinsed
   with water after use.

Specimen result

The breathed air tube contained much more water than the unbreathed air tube.

Think about it!

1   There is water in the air around us.
2   The water is in the form of a vapour.
3   Cooling turns water vapour into liquid water.

Suggested conclusion

Breathed air contains more water vapour than unbreathed air.

Method D: Heat production

Material per group/pupil

Thermometer

Specimen result

	Temperature
Unbreathed air	$20^{\circ}C$
Breathed air	$32^{\circ}C$

Suggested conclusion

Breathed air is warmer than unbreathed air.

Suggested final conclusion

It is clear that my cells are carrying out aerobic respiration because I am
absorbing oxygen from the air and producing carbon dioxide, water vapour and heat
energy.

4.3 Heat production in respiration

Object

To show that respiring tissues release heat energy

Method A: Teacher demonstration

Material per class

Small thermos flasks (2 per group)
Thermometers reading, if possible, to $0.2^{\circ}C$
Cotton-wool
Disinfectant, about 250 ml per group. This may be either 1% formalin solution
  or commercial bleach (sodium hypochlorite) diluted with 4 times its volume of
  water.
Soaked seeds, eg wheat or peas - enough, when dry, to half-fill each flask

Material per group/pupil

2 retort stands with clamps wide enough   Bunsen
  to take the flasks   Tripod/wire gauze
Beaker (for boiling seeds)   3 labels

Comments

If numbers are too large, just one experiment could be set up as a demonstration
and the results shared.
The vacuum flasks and thermometers should be rinsed with disinfectant before use.

Specimen result

Flask	Seeds	Temperature			
		Start	1 day	2 days	3 days
A	Live	$18.6^{\circ}C$	$20.2^{\circ}C$	$23.6^{\circ}C$	$24.0^{\circ}C$
B	Killed	$18.6^{\circ}C$	$19.2^{\circ}C$	$18.0^{\circ}C$	$19.0^{\circ}C$

Method B: Pupil activity

Material per class

Differential air thermometers, as many as possible
Various living organisms, one type for each thermometer, eg maggots, woodlice,
   mealworms, locust nymphs, earthworms, small mammal, fresh living plant material,
   in sufficient quantity to cover the bottom of the test chamber at least.

Specimen result

None

Think about it!

1   Food contains chemical energy.
2   Organisms can convert this to kinetic, heat, sound, light and electric
     energy.
3   The process by which organisms do these energy changes is called respiration.
4   The type of respiration being observed is aerobic respiration.
5   Heat energy is always released in respiration.

Suggested conclusion

When living organisms release energy from their food in respiration, some is
always released as heat energy.

4.4   Alcoholic fermentation

Method A: Brewing and yeast

Object

To show that, in the absence of oxygen, yeast can carry out respiration, producing
carbon dioxide and alcohol

Material per class

10% glucose - allow 30 ml per group, boiled, allowed to cool and kept in a sealed bottle.
30 ml air equilibrated bicarbonate indicator (see page 7) per group. This can be collected at the end, aerated again and re-used.
Liquid paraffin.

Material per group/pupil

2 boiling tubes with delivery tubes and stoppers as shown in the <u>Experiment Guide</u>
2 pieces cotton-wool, cut to fit tubes, with elastic bands to secure
Test tube rack

Specimen result

	Temperature	Indicator colour	Alcohol present?
Test tube A	22°C	Yellow	Yes
Test tube B	20°C	Red	No

Think about it!

1 The glucose solution was first boiled to drive out air and the mixture covered with oil to keep out air.
2 Heat energy is a sign that respiration is taking place.
3 The cotton-wool keeps in the heat.
4 The energy comes from the glucose.
5 The indicator shows that carbon dioxide has been given off.
6 Lime water can also be used to detect carbon dioxide.
7 The control test tube shows that the yeast is causing the changes.

Suggested conclusion

In the absence of air, yeast can obtain energy from glucose (sugar), which it converts to alcohol and carbon dioxide.

Method B: Baking and yeast

Object

To show how anaerobic respiration in yeast is made use of in bread-making

Material per class

Yeast suspension. Mix 30 g of dried yeast and 5 g of sugar with 150 ml of warm water. Leave in a warm place until visibly bubbling. (Fresh yeast from a baker works faster.)
Plain flour
Sugar
Scoops from tins of babyfood

Material per group/pupil

```
2 100 ml beakers
Plastic spatulas or stirring rods
Labels
```

Comments

```
If time presses, the beakers of dough may be left overnight, then kept in a fridge
if necessary.  The dough samples from all groups could be combined for baking or
special samples made up to demonstrate the difference.
```

Specimen result

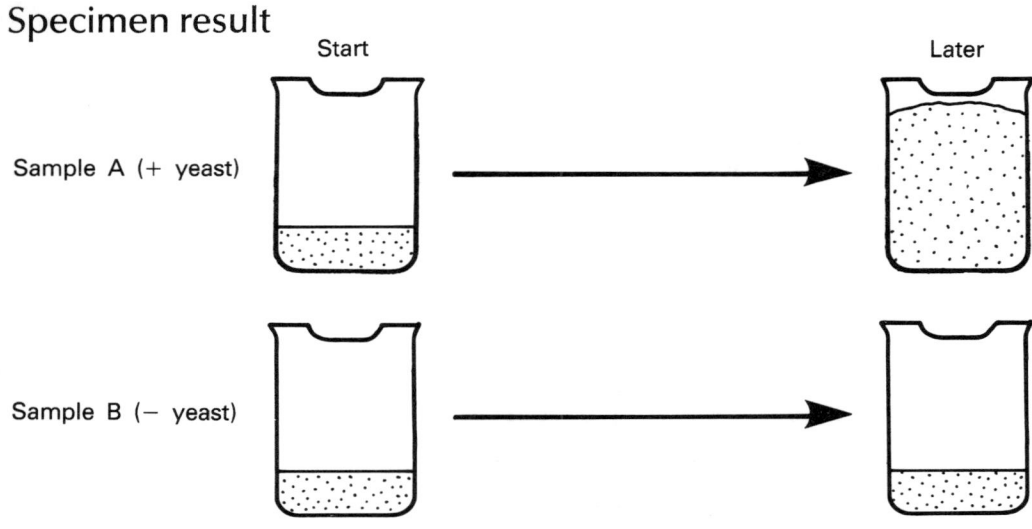

Dough	Texture of bread	Appearance of bread
A + yeast	Soft /~~Hard~~	Spongy
B − yeast	~~Soft~~/Hard	Solid

Think about it!

```
1    Carbon dioxide could make the dough rise.
2    Carbon dioxide is produced by anaerobic respiration by the yeast.
3    Alcohol is also formed.
4    During baking, the alcohol boils off.
5    Yeast makes bread spongy and soft.
```

Suggested conclusion

```
The carbon dioxide given off by the anaerobic respiration of yeast during bread-
making bubbles through the dough to give the bread a light, spongy texture.
```

4.5   Organisms and the atmosphere

Method A: Organisms and carbon dioxide

Object

To show that a wide variety of different organisms give off carbon dioxide

Material per class

Air-equilibrated bicarbonate indicator (see page 7)
A variety of plant and animal specimens
Appropriate sealed containers for organisms

Specimen result

Organism	Bicarbonate Indicator	
	Colour at start	Colour at finish
Pond snail	Red	Yellow
Locust	Red	Yellow
Earthworm	Red	Yellow
Elodea in light	Red	Purple
Fish	Red	Yellow
Germinating seeds	Red	Yellow
Elodea in dark	Red	Yellow
Small mammal	Red	Yellow

Suggested conclusion

All organisms seem to give off carbon dioxide, though, in light, plants take in more than they give out.

Experiment B: Gas absorption by small organisms

Object

To show that various small organisms absorb gas, presumably oxygen, from the air.

Material per class

3 boiling tubes fitted with stoppers, with lengths of
  capillary tubes and wire gauze barrier
Dropper with ink
Soda-lime
Retort stand with clamps to hold tubes one above the other
Suitable small organisms, eg locusts, maggots, mealworms,
  earthworms, germinating seeds

Specimen result

Apart from the control which should not move, the ink drops will move inward as gas
is absorbed.

Think about it!

1   If the organisms absorb gas, the ink drops should move inwards.
2   If the organisms give out carbon dioxide gas, this should push the ink drops
     outwards.
3   The soda-lime absorbs any carbon dioxide given out.
4   If the tube got cooler, the ink drop would move inward because cooling a gas
     makes it contract (get smaller).
5   If temperature affected the experiment, the ink drop in the control would move.

Suggested conclusion

The organisms tested were absorbing gas from the air in the tube. The gas is, in
fact, oxygen.

4.6   Gas analysis: inhaled and exhaled air

Object

To measure the percentages of carbon dioxide and oxygen in samples of inhaled and
exhaled air

Material per class

Saturated potassium hydroxide solution. Add about 50 ml of water to 100 g of
  potassium hydroxide in a hard glass flask, mixing continuously. Keep the flask
  cool, when mixing, by swirling under the tap. Store in Polystop dispensing
  reagent bottles.
Potassium pyrogallate solution is made up freshly in small amounts when required
  as follows:
  Put roughly 1 volume of potassium hydroxide pellets and 5 volumes of resublimed
  pyrogallol in a 100 ml beaker. Simultaneously, add 2 volumes of water and 1
  volume of liquid paraffin and gently stir the lower water layer until the solid
  dissolves.

It should be possible to make a clear solution (potassium pyrogallate goes black in contact with oxygen).
It must be stressed to pupils that both these reagents are very caustic and must be handled with great care. 2M hydrochloric acid coloured with litmus solution should be kept handy to neutralise spillages.

Material per group/pupil

Labelled 100 ml beaker containing a small amount of saturated potassium hydroxide solution
Labelled 100 ml beaker containing a small amount of potassium pyrogallate solution prepared as above
Ruler
Paper towels to protect bench
Clean J-tube. After each experiment the J-tube should be flushed with dilute acid, then hot detergent, rinsed out with water and dried with air.

Comments

If there is enough equipment, the experiments can be done step by step as a class under the teacher's direction.
The technique of use can be demonstrated by filling the J-tube with coloured water and laying it flat on an overhead projector.
One tube of breathed air supported over a trough of water will supply the whole class.

Specimen result

Unbreathed air

Original length of column $= 109\,\text{mm}$

Length after hydroxide $= 108\,\text{mm}$

Length due to CO_2 = 109-108 $= 1\,\text{mm}$

Percentage $CO_2 = \dfrac{109-108 \times 100}{109}$ $= \underline{0.92\%}$

Length after pyrogallate $= 87\,\text{mm}$

Length due to O_2 = 108-87 $= 21\,\text{mm}$

Percentage $O_2 = \dfrac{108-87 \times 100}{109}$ $= \underline{19.3\%}$

Breathed air

Original length of column $= 115\,\text{mm}$

Length after hydroxide $= 111\,\text{mm}$

Length due to CO_2 = 115-111 $= 4\,\text{mm}$

Percentage $CO_2 = \dfrac{115-111 \times 100}{115}$ $= \underline{3.5\%}$

Length after pyrogallate $= 92\,\text{mm}$

Length due to O_2 = 111-92 $= 19\,\text{mm}$

Percentage $O_2 = \dfrac{111-92 \times 100}{115}$ $= \underline{16.5\%}$

Suggested conclusion

In breathing, some oxygen is removed from the air and replaced with a similar amount of carbon dioxide.

4.7   The effect of exercise on breathing

Object

To show the effect of exercise on the rate of breathing, the depth of breathing, and the carbon dioxide content of exhaled breath.

Material per class

Air-equilibrated bicarbonate indicator (see page 7)
Labelled beaker for used indicator. (This can be aerated again and re-used.)
20 ml syringes or measuring cylinders

Material per group/pupil

2 boiling tubes   Drinking straws
Test tube rack   Stopclock

Specimen result

1

	Before exercise	After exercise
1st count	21	42
2nd count	19	34
3rd count	22	27
Average	21	34

2   The size of breaths I took after exercise was smaller/the same/larger than before exercise.
3   The time it took for my exhaled breath to turn 20 ml of bicarbonate indicator from red to yellow was:
         Before exercise : 40 s
         After exercise   : 20 s

Think about it!

1   During exercise, we must have more oxygen.
2   Oxygen is needed for the muscles to produce more energy.
3   I obtained more oxygen by breathing faster and more deeply.
4   During respiration, carbon dioxide is produced.
5   Carbon dioxide is breathed out.
6   Carbon dioxide turns bicarbonate indicator from red to yellow.

Suggested conclusion

During exercise, muscles need more oxygen to provide more energy. To get this, we breathe faster and deeper. This also helps remove the extra carbon dioxide produced.

4.8 Breathing models

Method A: Diaphragm action

Object

To represent the effect of diaphragm movements on the lungs

Material per class

One bell jar diaphragm model (Joins must be sealed securely and the rubber sheet treated very gently.)

Specimen result

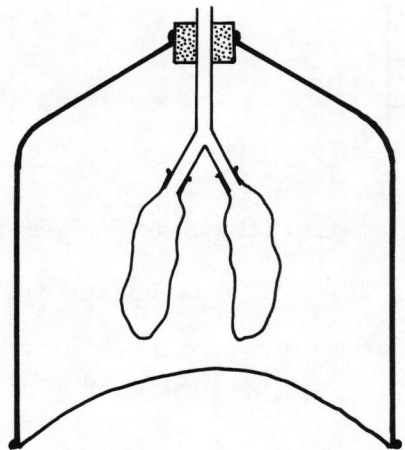 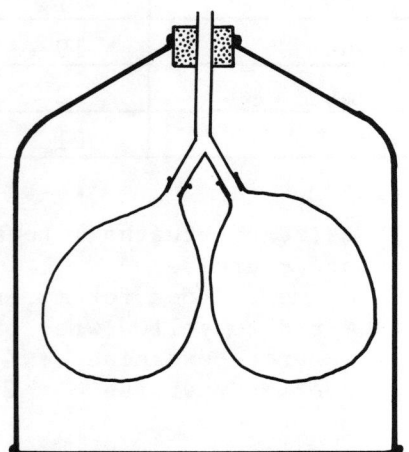

Suggested conclusion

When the diaphragm contracts, it becomes ~~curved~~/straight and moves downwards/ ~~upwards~~. This ~~decreases~~/increases the volume of the thorax causing the lungs to ~~empty~~/fill.

Method B: Intercostal muscle action

Object

To show the relative lengths of the external and internal intercostal muscles when the sternum is raised or lowered

Material per class

A model, approx 25 x 15 cm, made from strong wood securely rivetted together. This should be firmly clamped by the 'spine' in a heavy stand. Each part, including the muscles, should be labelled clearly on the model.

Specimen result

Position of sternum	External intercostals (contracted or relaxed?)	Internal intercostals (contracted or relaxed?)
Raised	Contracted	Relaxed
Lowered	Relaxed	Contracted

Suggested conclusion ·

When the external intercostal muscles contract, they ~~lower~~/raise the sternum and ribs, increasing/~~decreasing~~ the volume of the thorax, causing the lungs to ~~empty~~/fill.

Think about it!

1   The bell jar is not a good representation of the ribcage because it does not move.
2   The balloons are not an accurate representation of lung structure since lungs actually have millions of air bags (alveoli) not just two.
3(a) If the model trachea was blocked, the ballons could not fill or empty.
 (b) If the rubber sheet had a hole in it, the balloons would not fill or empty properly.
4   The heart is also found inside the thorax.
5   The elastic bands are not really a good representation of the intercostal muscles since they are under tension when they are long rather than short like muscles.

4.9   Vital capacity and tidal volume of the lungs

Object

To estimate the vital capacity and tidal volume of the lungs

Material per class

Large plastic container with screw cap or stopper. A capacity of at least 5 l is desirable. This should be clearly graduated at 100 ml intervals.
About 1 m of plastic or rubber tubing. After each experiment, this should be rinsed in disinfectant, eg a 1% solution of an ampholytic disinfectant, then washed with water.

Specimen result

Vital capacity   = 4600 ml
Tidal capacity   (1)   = 560 ml
                  (2)   = 490 ml
                  (3)   = 520 ml

$$\text{Average} = \frac{\text{Total}}{3} = \frac{1570}{3} = 523 \text{ ml}$$

Think about it!

1   The body needs oxygen to produce energy in respiration.
2   During exercise, the body needs more oxygen.
3   The larger vital capacity provides for extra needs during exercise.

Suggested conclusion

The vital capacity is much greater than the tidal volume so that, when we use more
energy, eg in exercise, there is spare capacity in the lungs to absorb more oxygen.

4.10 Examining stomata

Object

To examine the structure and distribution of stomata

Method A: Structure

Material per group/pupil

Fresh lettuce leaf   Mounted needle
Petri dish   Dropper and water
Scalpel   Microscope, with bench lamp
Microscope slide   if necessary
Cover slip

Specimen result

None

Think about it!

1   Guard cells are sausage-shaped, have chloroplasts and an unevenly-thickened wall.

> 2   Guard cells can carry out photosynthesis since they have chloroplasts and other epidermal cells do not.
> 3   If the guard cells straighten, the stoma is closed.

Method B: Distribution

Material per group/pupil

Dicotyledon leaf, eg privet, <u>Tradescantia</u>   Cover slip
Clear nail varnish   Mounted needle
Forceps   Dropper and water
Microscope slide   Microscope and bench lamp, if necessary

Specimen result

<u>No of stomata seen</u> (x 75 privet)

Upper epidermis : 0
Lower epidermis : 9

Think about it!

1   Water vapour could also escape through the stomata.
2   The plant could lose too much water and wilt.
3   The water loss problem would be worse on a hot summer's day.
4   The upper surface of a leaf will be most heated by the sun.
5   There are many more stomata on the lower surface of the leaf. Since the lower surface gets less heat from the sun than the upper, this helps cut down water loss.

Suggested conclusion

A stoma is a hole between 2 guard cells. Guard cells are sausage-shaped, with chloroplasts and an unevenly-thickened cell wall, unlike other epidermal cells. There are also more stomata on the lower than on the upper epidermis. This helps cut down water loss.

5.1   The water content of food

Object

To show that most foods consist mainly of water

Material per class

Oven set at 80-90°C
Balance(s), capable of weighing to at least 0.1g
A variety, as many as possible, of fresh foods

Material per group/pupil

Porcelain basin (75 ml) or other suitable
  dish
Knife or scalpel

Bench mat
Label

Specimen result

Food	Percentage of water
Lettuce	95%
Meat	63%
Fish	77%
Beans	60%
Apple	87%
Bread	36%
Cabbage	94%

Suggested conclusion

Most foods consist mainly of water.

5.2   Testing foods for carbohydrates, proteins and fat

Object

To analyse foods for the presence of sugar, starch, protein and fat

Material per class

A variety of foods. These might include animal flesh, offal, dairy products, eggs,
  flour products, cereals, pulses, nuts, tubers and roots, margarine, fruits.
The required chemicals, preferably in small labelled dropping bottles supplied to
  each group. These are:

    Benedict's solution ⎫
    Iodine solution   ⎬ see page 6
    5% sodium hydroxide solution

    1% copper sulphate solution
    Propan-2-ol
    2 M hydrochloric acid

Material per group/pupil

Mortar and pestle
Small beaker (100 ml)

Filter funnel
Labels

Spotting tile
3 test tubes
Test tube rack
Filter papers

Large beaker (400 ml)
Beaker
Tripod/wire gauze
Bunsen (or thermostatic water bath set
    at 80-90°C)

Specimen result

None

Suggested conclusion

General conclusions about the accuracy of the tests or the contents of particular
food types, eg meats, fruits, pulses, etc.

5.3   Measuring vitamin C concentration

Object

To measure the vitamin C concentration of a range of fruit juices

Material per class

Freshly-prepared 0.1% DCPIP solution - 6 ml per group
0.1% vitamin C solution prepared by crushing a 50 mg tablet in a mortar and pestle,
    adding 50 ml water a little at a time and decanting each portion through coarse filter
    paper to remove chalk or starch
Various lightly-coloured fruit juices
1 ml syringes or pipettes to dispense DCPIP

Material per group/pupil

Test tube and rack
2 ml syringe with needle or graduated pipette

Comments

The instruction to 'stir not shake' is to prevent recolouring of the DCPIP.
If a syringe is used, the needle and test tube must be of a size that makes stirring
possible.

Specimen result

Juice tested	Volume needed	Vitamin C concentration
Standard Vit C	0.4 ml	0.1%
Lemon juice	0.5 ml	0.08%

Suggested conclusion

Fruit juices are good sources of vitamin C.

5.4   The effect of cooking on vitamin C

Object

To show that while vitamin C is reduced somewhat by heating, vitamin C loss is much greater when it is dissolved out of a food during boiling.

Material per class

Freshly-prepared 0.1% DCPIP solution - 6 ml per group
1 ml syringes or pipettes to dispense DCPIP
A few fresh oranges
Domestic orange squeezer
Step 1: beaker, filter funnel, glass wool
Step 2: beaker, graduated beaker, tripod/wire gauze, bunsen
Step 3: beaker, tripod/wire gauze, bunsen
Step 4: beaker, filter funnel, glass wool

Note: Steps 1 to 4 could be allocated to different groups and the resulting liquids shared out for Step 5.

Material per group/pupil

Test tube and rack
2 ml syringe with needle or graduated pipette

Comments

Prolonged heating of a food will eventually have a drastic effect on its vitamin C content. It might be more difficult to detect here. Make sure that in Step 2 at least all the water boiled off is replaced.

Specimen result

Juice tested	Volume needed	Vitamin C concentration
Standard vitamin C	0.4 ml	0.1%
Fresh orange juice	0.7 ml	0.06%
Heated orange juice	0.8 ml	0.05%
Boiled orange juice	0.9 ml	0.04%
Cooking water	1.8 ml	0.02%

Think about it!

1   Heating does not completely destroy vitamin C.
2   Heating reduces vitamin C a little.
3   Boiling a food reduces its vitamin C a lot.
4   It is not the heat; the vitamin C is dissolved out of the food by the water.
5   If we used the cooking water to make soups or gravy, we would not waste the vitamin C.

Suggested conclusion

Heat can destroy vitamin C if it is too prolonged. However, since it dissolves in water it is easily lost when vegetables are boiled. The cooking water could be used for soups or gravy to prevent this.

5.5   The energy in a peanut

Object

To estimate the calorific value of a peanut

Material per class

Balance, as accurate as possible
10 ml syringes or measuring cylinders

Material per group/pupil

Shelled peanuts   Thermometer
Boiling tube   Mounted needle or long pin
Retort stand and clamp   Bunsen

Specimen result

```
Weight of peanut            = 0.52 g
Temperature of water at finish = 53 °C
Temperature of water at start = 22 °C
Energy content of peanut    = 1.19 Cal/g = 5.01 kJ/g
```

Think about it!

1   A peanut contains chemical energy.
2   It is converted mainly into heat energy but also into light, kinetic and sound energy.
3   It is the heat energy which is actually measured.
4   Not all the peanut's energy goes into the water. Some escapes round the sides of the test tube.
5   The water is stirred to make the temperature of the water the same all over.
6   Because of heat losses, the result would be too low.

Suggested conclusion

Peanuts contain a lot of energy. This simple experiment allows too much heat to escape to be accurate. The result is probably too low.

5.6   The energy content of food

Object

To demonstrate a more accurate method of measuring the calorific value of foods

Material per class

```
Food calorimeter (from Griffin or Philip Harris)
Oxygen cylinder with delivery tube
Thermometer reading to 0.1 °C
Power pack set at 6 V
Balance, as accurate as possible, to weigh food
Pure carbohydrate, eg powdered starch or sugar
Pure protein, eg albumen, flake
Steak, about 2 g, weighed, chopped into a crucible, and dried
Measuring cylinder, eg 250 ml
```

Comments

Although the instructions are written as if for pupils, this is a demonstration. Pupils can be enlisted to measure the water into the calorimeter, stir and take the temperatures.

Specimen result

<u>Carbohydrate</u>
Weight of water = 616 g
Weight of carbohydrate = 1.12 g
Temperature of water at finish = 25°C
Temperature of water at start = 18.3°C
Energy content of pure carbohydrate = 3.69 kCal/g = 15.48 kJ/g

<u>Protein</u>
Weight of water = 621 g
Weight of protein = 0.97 g
Temperature of water at finish = 24.5°C
Temperature of water at start = 18.4°C
Energy content of pure protein = 3.89 kCal/g = 16.34 kJ/g

<u>Steak</u>
Weight of water = 626 g
Weight of steak (fresh) = 1.96 g
Temperature of water at finish = 26.7°C
Temperature of water at start = 18.1°C
Energy content of steak = 2.74 kCal/g = 11.5 kJ/g

Think about it!

1   Oxygen is allowed to flow for a while before ignition to flush air out of the system.
2   Foods burn more quickly in oxygen.
3   The stirrer must be used to make sure that the heat is spread evenly throughout the water.
4   The copper tube transfers heat quickly from the burning chamber into the food.
5   Heat may be lost through the sides and the apparatus itself absorbs some heat.
6   The igniter coil allows the food to be lit without removing the water.
7   The final temperature must be taken quickly before heat escapes.

Suggested conclusion

The calorific value of the whole food is lower than that of the pure nutrients since the food also contains water.

6.1   Feeding adaptations

Object

To examine the feeding methods of <u>Amoeba</u>, mussel and housefly.

Material per class

<u>Amoeba proteus</u> cultures can be obtained from Philip Harris and Griffin with full

details for observation and culture.

Live mussels (<u>Mytilus</u>) with sea water, either 'genuine' or made up with sea salt from a supplier. Freshwater mussels may also be used and, for part of the investigation, pickled mussels can be used.

Tethered flies - blowfly (<u>Calliphora</u>) larvae can be obtained from suppliers, pet shops, etc. These are put, with their food, on to a tub of sawdust inside a suitable small cage. Kept in a warm place, they will pupate and hatch. To prepare them, the cage is put in a freezer for 5 minutes. This anaesthetises the flies. Melt some candle wax over a hot plate. Take a piece of thread about 30 cm long. Roll a little ball at one end and dip it into the wax. Quickly place this on the back (thorax) of an anaesthetised fly keeping it away from the eyes and wings. Hold it steady until the wax hardens.

Sugar solution.

Material per group/pupil

Microscope slides
Cover slips
Mounted needle

Ink or carmine with dropper
Scalpel (blunt)

Specimen result

None

Suggested conclusion

Animals show many feeding adaptations, depending on their habitat and food.

6.2 The effect of acid on teeth

Object

To show that even dilute acids can dissolve tooth enamel

Material per class

Teeth, ideally 3 per group. Human teeth may be obtained from a friendly local dentist. Animal teeth, preferably incisors, can be had from butchers or abattoir. If teeth are awkward to get, each pupil could prepare just one tooth or the experiment could be done as a demonstration.

2 M hydrochloric acid
Fizzy cola

Material per group/pupil

3 test tubes
Test tube rack
Labels
Scalpel

Coloured nail varnish
Tweezers
Paper towels

Specimen result

Tooth	A	B	C
Bathing liquid	Water	Dilute acid	Cola drink
pH	7	1	2-3
Appearance at start			
Appearance at finish			

Description of teeth : The tooth in water was unaffected. The tooth in acid had the letter cut deeply into it. The letter was also cut by the cola but not so deeply.

Think about it!

1   Acid dissolves tooth enamel.
2   Test tube A shows that acid, not just liquid, dissolves teeth.
3   Acid in the mouth is produced by bacteria (plaque) which feed on sugar in the mouth.
4   Fizzy drinks and other foods could cause harm because they contain acid.
5   Toothpastes should have a high pH. Low pHs are acid. pHs above 7 are alkali. Alkali in toothpaste will help neutralise acid in the mouth.

Suggested conclusion

Acid dissolves tooth enamel. Although this is normally caused by bacteria (plaque) in the mouth which convert sugar to acid, some foods and drinks are also acid.

6.3   Toothpastes

Object

To compare the pH and abrasive content of a variety of popular toothpastes

Material per class

Tubes of toothpaste, various popular brands
Dropping bottles containing 10% suspension of the same toothpastes. (The bottles can
  be labelled effectively by taping the names from the packets around them.)
Water with droppers

Material per group/pupil

pH papers
pH colour chart
Microscope slides
Cover slips

Mounted needle
Microscope, with bench lamp
  if necessary

Specimen result

None

Think about it!

1   If the local water supply contains fluoride, we may not want it in our
     toothpaste as well. Also, some people think it could be harmful.
2   pH values below 7 are acid and above 7 are alkali.
3   pH values of above 7 in toothpaste should help resist tooth decay since
     this would help to neutralise mouth acid.
4   The grit in toothpaste could help scrape plaque off teeth.

Comments

Most toothpastes contain abrasives such as calcium carbonate, not only to remove
plaque but also for the cosmetic purpose of polishing the teeth. They can, however,
also cause damage to the surface of the teeth. This depends largely on the tooth-
brushing habits of the individual. A conscientious toothbrusher would benefit from
a toothpaste with a fine abrasive (the situation assumed in this experiment), while
an irregular brusher might need a coarser abrasive to achieve a worthwhile
cleaning effect.

Suggested conclusion

Toothpastes with an alkaline pH (above 7) may help neutralise mouth acid and so
reduce decay. Fluoride in the toothpaste will also help. The grit toothpastes
contain might help in scraping off plaque. Of the toothpastes we tested, I think
the best one would be ...

6.4   The need for digestion

Object

To show that glucose molecules are small enough to pass through the (model) gut wall
but starch molecules are not

Material per class

Starch - glucose solution (see page 6)
Funnel or syringe to fill bags

Material per group/pupil

Visking tubing (14 mm diameter),
  15-20 cm length
Thread
Boiling tube
Test tube
Test tube rack
Spotting tile
Large beaker
Tripod/wire gauze } or thermostatic
Bunsen   } water bath set at 80°C

Labels
Dropper
Benedict's solution in labelled
  dropping bottle (see page 6)
Iodine solution in labelled
  dropping bottle (see page 6)

Comments

If time is short, the tubing can be left until the next lesson before the final sample is taken.

Specimen result

Time (min)	0	10	20	30	40	50
Glucose	0	+	++	+++	++++	+++++
Starch	0	0	0	0	0	0

Think about it!

1   Only glucose leaked from the bag.
2   The Visking tubing must have tiny holes in it to allow things to get through it.
3   Starch molecules must be bigger than glucose molecules since glucose got through the holes but starch could not.
4   Starch molecules are made of long chains of hundreds of glucose molecules joined together.
5   Only glucose molecules could get from the inside of the gut to the rest of the body. Intact starch molecules would be too large.
6   If starch molecules were chopped up into glucose, the body could then get the benefit of them.
7   Proteins and fats are other large food molecules which might need chopping up before the body could use them.

Suggested conclusion

Large insoluble food molecules like starch cannot pass through the gut wall to reach the rest of the body. First they must be broken down into smaller soluble molecules like glucose. This breaking down is called digestion.

6.5 Digestion in a model gut

Object

To represent digestion and absorption of food in a model gut

Material per class

2% starch solution with 0.1% sodium chloride (see page 6)
Water baths at $37^{o}C$ and $80^{o}C$ if available
2 funnels or syringes to fill bags

Material per group/pupil

Visking tubing (14 mm diameter)
  15-20 cm length
Thread
Small beaker for saliva
Boiling tube
Test tube
Test tube rack
Spotting tile
Large beaker
Tripod/wire gauze } or thermostatic
Bunsen   } water bath set at $80^{o}C$

Labels
Dropper
Benedict's solution in labelled
  dropping bottle (see page 6)
Iodine solution in labelled
  dropping bottle (see page 6)

Comments

If possible, the boiling tubes and contents should be kept in a water bath at $37^{o}C$ throughout.

Specimen result

Time (min)	0	10	20	30	40	50
Sugar	0	0	+	++	+++	++++
Starch	0	0	0	0	0	0

Think about it!

1   The Visking tubing must have tiny holes in it to allow things to get through it.
2   No starch molecules escaped from the bag. We know this because the tests with iodine were all negative.
3   Starch molecules are too big to escape through the tiny holes in the bag.

4   Starch molecules are made of long chains of hundreds of glucose molecules joined together.
5   Sugar did leak out of the bag during the experiment. We know this because the tests with Benedict's solution became positive.
6   There was no sugar in the bag at the start.
7   The sugar could have come from the starch molecules breaking down into smaller molecules.
8   The saliva contains an enzyme called amylase which can break up starch molecules.
9   Digestive enzymes like amylase break down large insoluble food molecules like starch, proteins and fats into smaller soluble molecules which can pass through the gut wall to reach the rest of the body.

Suggested conclusion

Digestive enzymes (like the amylase in saliva) break down large insoluble food molecules (like starch) into smaller soluble molecules (like sugar) which can pass through the walls of the digestive system to supply the rest of the body.

6.6   The effect of saliva on starch

Object

To show that saliva contains an enzyme which can convert starch into sugar (maltose)

Material per class

1% starch solution with 0.1% sodium chloride (see page 6)

Material per group/pupil

Small beaker for saliva
5 ml syringe
Spotting tile
Test tube
Test tube rack
Large beaker
Tripod/wire gauze } or thermostatic water
Bunsen   } bath set at 80°C

Iodine solution in labelled
   dropping bottle (see page 6)
Benedict's solution in labelled
   dropping bottle (see page 6)
Stopclock

Comments

Good mixing is essential throughout this experiment.

Specimen result

Time (min)	0	$\frac{1}{2}$	1	$1\frac{1}{2}$	2	$2\frac{1}{2}$	3	$3\frac{1}{2}$	4	$4\frac{1}{2}$	5
Iodine test											
Time (min)	$5\frac{1}{2}$	6	$6\frac{1}{2}$	7	$7\frac{1}{2}$	8	$8\frac{1}{2}$	9	$9\frac{1}{2}$	10	$10\frac{1}{2}$
Iodine test											

The test with Benedict's was positive/~~negative~~.

Think about it!

1   Starch was present at the start but not at the end of the experiment.
2   We could have heated some saliva with Benedict's solution.
3   Sugar was present at the end of this experiment.
4   Starch molecules are made of sugar (glucose) molecules joined together.
5   The large starch molecules seem to have been chopped into smaller sugar molecules.
6   The smaller molecules are able to leak out of the gut to reach the rest of the body.
7   You cannot say for sure that the saliva caused the change. To check, you could set up a tube containing starch and water instead of starch and saliva.
8   If the saliva had been boiled, nothing would have happened since heat denatures enzymes so that they cannot work.

Suggested conclusion

Saliva contains an enzyme which breaks down large insoluble starch molecules into small, soluble sugar (maltose) molecules. This enzyme is called salivary amylase (or ptyalin).

6.7   The effect of pepsin on protein

Object

To show that pepsin digests protein but only in acid conditions

Material per class

1% pepsin solution - 3 ml per group for experiment A and 30 ml per group for B
Dilute hydrochloric acid in labelled dropping bottle (2 M)
Dilute potassium hydroxide in labelled dropping bottle
Thermostatic water bath set at $37^{\circ}C$
Experiment A: egg albumen suspension, 40 ml per group (see page 7)

Experiment B: hard boiled egg white cut into 1 cm³ cubes, 4 per group
1 ml and 10 ml syringes to dispense liquids
Beaker of water

Material per group/pupil

4 test tubes
Test tube rack
Labels

Large beaker and thermometer
 (or water bath at 37°C)

Method A: Specimen result

Test tube	Cloudy/Clear
A	Clear
B	Cloudy
C	Cloudy
D	Cloudy

Method B: Specimen result

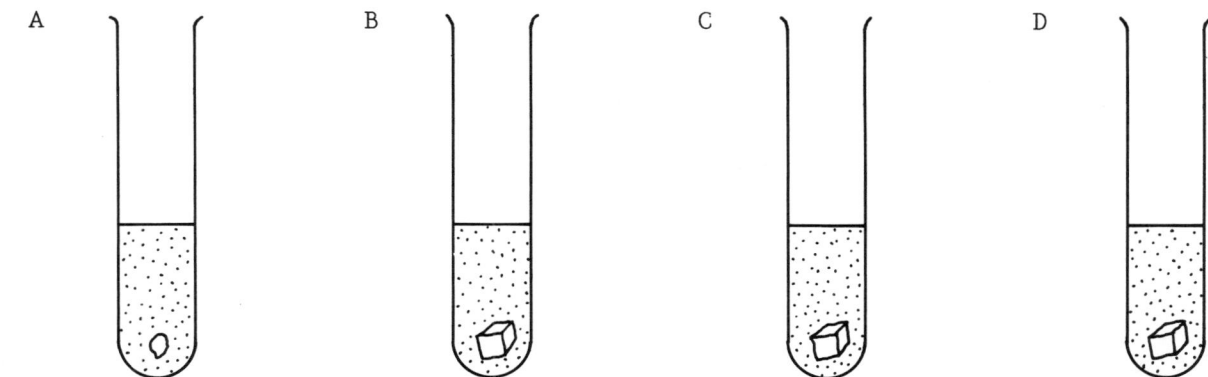

Think about it!

1   Protein has been digested only in test tube A.
2   Pepsin has digested the protein. In test tube B, which contains no pepsin,
     protein was not digested.
3   Pepsin needs acid conditions in which to work. Pepsin has to work in the acid
     conditions of the stomach.
4   Hydrochloric acid does not digest proteins. Test tube B had hydrochloric acid
     alone and digestion did not occur.
5   Proteins are made of smaller molecules called amino acids.
6   The body digests proteins to break down the large, insoluble protein molecules
     into small, soluble amino acids which are able to leak out of the gut to reach
     the rest of the body.
7   The test tubes are incubated at 37°C because this is the temperature of the
     body.

Suggested conclusion

Pepsin can break down (digest) large, insoluble protein molecules into smaller, soluble molecules. It only works in acid conditions (low optimum pH) such as those found in the stomach.

6.8   The effect of lipase on fat

Object

To show that lipase causes (fatty) acids to be produced from milk fats and that bile salts help this process

Material per class

Beaker of fresh milk, 15 ml per group
Beaker of freshly-prepared 5% lipase solution, 10 ml per group
Beaker of freshly-prepared 1% bile salts solution, 2 ml per group
Beaker of distilled water
5 ml and 1 ml syringes to dispense the above
0.01% phenol red solution in labelled dropping bottles (litmus solution or
   phenolphthalein are alternatives)
Dilute sodium hydroxide in labelled dropping bottles
Thermostatic water bath set at 37°C

Material per group/pupil

3 test tubes   Large beaker (or water bath at 37°C)
Test tube rack   Labels

Specimen result

Test tube \ Time	0	5	10	15	20	25	30
A							
B							
C							

Think about it!

1   A fat molecule consists of one glycerol molecule with 3 fatty acids joined to
     it.
2   The acid in the test tube could come from fatty acids being split from the
     fat molecules.

3   The lipase could have split fatty acids from the fat molecules.
4   Lipase did cause this to happen. Test tube C had no lipase and acid did not form.
5   Bile salts could not cause this to happen. Test tube C did have bile salts and acid did not form.
6   Bile salts speeded up the action of the lipase.
7   The test tubes were incubated at 37°C because this is the temperature of the body.
8   Water was added to make sure all the test tubes contained the same volume and concentration of liquid.

Suggested conclusion

The lipase seems to be releasing fatty acids from the fat molecules. Bile salts help this process.

6.9   Bile salts

Object

To show that bile salts cause fats to emulsify

Material per class

Butter or margarine
Beaker of freshly-prepared 1% bile salts solution, 10 ml per group
Beaker of water
Thermostatic water bath set at 37°C
2 10 ml syringes

Material per group/pupil

2 test tubes   Glass rod
Test tube rack   Labels

Specimen result

None

Think about it!

1   The temperature of the human body is 37°C.
2   Inside the body, butter will melt.
3   When fats like butter are mixed with water, they quickly separate again.
4   Bile salts stop them separating - they form an emulsion.
5   The lipases which digest the fats are in watery juices. The bile lets them mix with the fats.

Suggested conclusion

Bile salts help fats to mix with water, forming an emulsion. This allows lipases in the watery digestive juices to get at them to digest them.

7.1 Water and organisms

Object

To show that organisms consist largely of water

Material per class

As wide a range of plant and animal tissues as possible, eg grass, privet stems, pot plants, pondweed, moss, ferns, seaweed, meat, bone (crush first), liver, kidney, heart, sweetbread (pancreas or thymus), fish, etc
All other material and equipment as Experiment 5.1

Specimen result

All tissues should be more than 50% water.

Suggested conclusion

Organisms consist mainly of water.

7.2 Sweat pores

Object

To indicate the existence of discrete sweat pores in the skin and to show that their number varies in different parts of the body

Material per group/pupil

Iodine solution in a labelled dropping bottle (see page 6)
Starch and Potassium Iodine Test Papers from Philip Harris or Griffin
Clear adhesive tape

Specimen result

Sweat glands are more numerous in some parts of the body, eg under the arms than in other parts.

Think about it!

1   The water came from the skin.
2   Sweat is made by sweat glands in the skin.
3   Sweat gets on to the skin through holes called sweat pores.
4   Sweat evaporates from the skin. To do this, it must absorb heat from the skin. This helps cool the body down.

Suggested conclusion

Sweat gets on to the skin through tiny pores. The number of pores varies in different parts of the skin.

7.3   Diffusion

Object

To demonstrate diffusion in gases and liquids and to show it is affected by the concentration gradient

Material per class

A few jars of 'brown gas'. The gas used is nitrogen dioxide prepared by carefully dropping concentrated nitric acid on to copper turnings in a gas jar inside a fume cupboard. This jar is then used to fill, by diffusion, the jars given to the pupils. The gas is toxic but the effect is rapid and gas jars can be quickly returned to the fume cupboard for disposal.
Gas jars with covers
Concentrated potassium permanganate solution
Dilute potassium permanganate solution made from the concentrated solution diluted to $\frac{1}{4}$ strength.

Material per group/pupil

2 test tubes
Test tube rack
Labels

Pasteur pipette with teat
Tissues

Specimen result

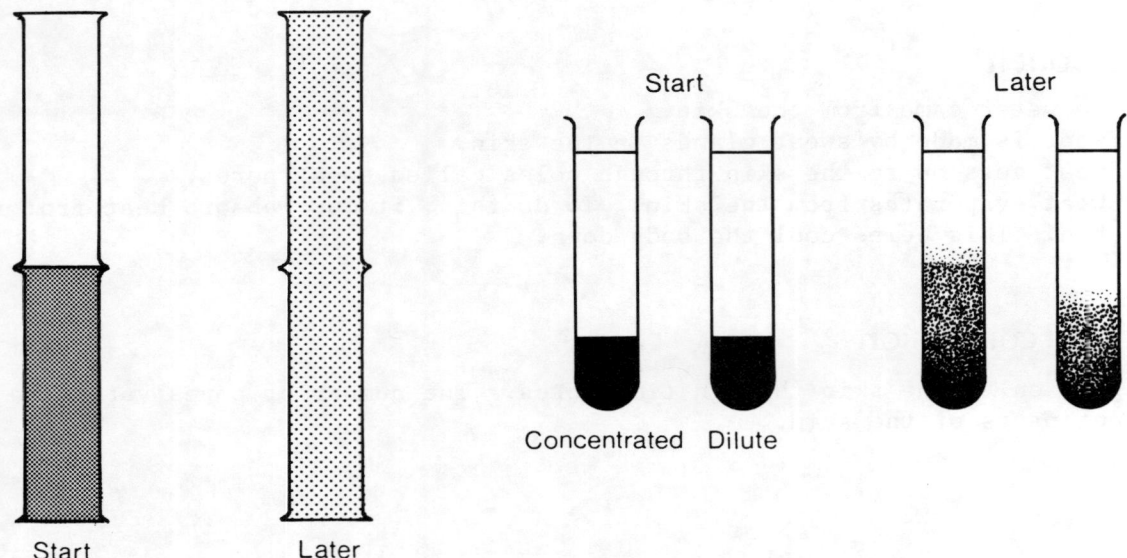

Start   Later

Start   Later

Concentrated   Dilute

Think about it!

1   The gas and permanganate particles were moving because we could see their colour spread.
2   The gas particles moved faster.
3   The concentrated permanganate test tube had the higher concentration gradient.
4   This did affect the rate of diffusion. The test tube with the higher concentration gradient diffused faster.

Suggested conclusion

Gas and liquid particles spread by diffusion. Gases diffuse faster than liquids. The higher the concentration gradient, the faster diffusion will take place.

7.4   Osmosis in living tissue

Object

To show osmosis taking place in living tissues

Method A: Potato

Material per group/pupil

3 potato halves
Scalpel or knife
3 Petri dish bases
Labels

Beaker
Tripod/wire gauze
Bunsen
Sugar

Specimen result

Water fills the cavity of A but not that of B or C.

Think about it!

1   The sucrose makes the water concentration in the cavity lower than in the surrounding cells.
2   In each dish, the water concentration is highest outside the potato.
3   Osmosis is the diffusion of water across a semi-permeable membrane from high water concentration to low water concentration.
4   The water moved into the cavity in B by osmosis. We know this because the water concentration there was lower than it was in the surrounding cells.
5   Semi-permeable membranes are needed for this to happen. We know this because it did not happen in A where the membranes had been destroyed by boiling.
6   There was no water concentration gradient in C so no osmosis took place.

Suggested conclusion

The sucrose crystals make the water concentration in the cavity lower than that around it. So water moves into the cavity by osmosis. This only occurs if the membranes are intact. Also, where there is no water concentration gradient (C), there is no movement of water by osmosis.

Method B: Grapes

Material per group/pupil

2 fresh, unblemished grapes cut from the bunch to leave a few mm of stalk intact
2 large boiling tubes
Test tube rack
Labels
Saturated salt solution

Specimen result

The grape in the salt floats and becomes softer. The other grape sinks and becomes firmer.

Comments

Leave longer than 48 hours if possible.

Think about it!

1 In A, the water concentration was higher outside the grape.
2 In B, the water concentration was higher inside the grape.
3 Osmosis is the diffusion of water across a semi-permeable membrane from high water concentration to low water concentration.
4 Grape B got lighter because it floated. We would expect it to get lighter because it loses water by osmosis.
5 Grape A got heavier because it sank. We would expect it to get heavier because it gains water by osmosis.

Suggested conclusion

In both cases, water moved from high water concentration to low water concentration. This caused grape A to swell and sink and B to shrink and float.

Method C: Shelled eggs

Material per class

2 fresh eggs with shells removed by immersing the eggs
   in dilute hydrochloric acid for a few hours
2 beakers
Labels
Saturated salt solution

Specimen result

The egg in water swells and sinks. The one in salt shrivels and floats.

Think about it!

1 In A, the water concentration is higher outside the egg.
2 In B, the water concentration is higher inside the egg.
3 Osmosis is the diffusion of water across a semi-permeable membrane from high water concentration to low water concentration.
4 Egg B got lighter because it floated. We would expect it to get lighter because it loses water by osmosis.
5 Egg A got heavier because it sank. We would expect it to get heavier because it gains water by osmosis.

Suggested conclusion

Same as method B

7.5   Osmosis in a model cell

Object

To demonstrate water movement by osmosis into and out of model cells

Material per class

Balance(s)
Tissues
20% sucrose solution,
   allow 200 ml per group

Beaker
Funnel or syringe to fill bags

Material per group/pupil

Visking tubing (14 mm diameter), 15 cm length
2 250 ml beakers
Labels

Specimen result

	Liquid inside	Liquid outside	Weight in grams		
			Start	30 min	60 min
A	20% sucrose	Water	16.73	20.50	25.18
B	Water	20% sucrose	15.76	11.80	8.38

Think about it!

1   In A, the higher water concentration is outside the bag.
2   The water should move into the bag. The bag should get heavier.
3   In B, the higher water concentration is inside the bag.
4   Water should move out of the bag. The bag should get lighter.
5   The bag would have got heavier but not as much.

Suggested conclusion

If the liquid outside has a lower water concentration than the (model) cell contents,
the cell will lose water (and weight) by osmosis. If the water concentration outside
is lower than that inside, water will enter the cell and it will gain weight.

7.6 The effect of heat and chemicals on the cell membrane

Object

To show that the cell membrane controls what enters and leaves cells and that it is destroyed by heat and certain chemicals

Material per class

Fresh beetroot
Cork borer, 8 mm
Test tube of ethanol, labelled
Test tube of concentrated acid
Test tube rack

Large beaker of crushed ice
Trough of water
Thermostatic water baths
    set at 40°C, 60°C, 80°C

Material per group/pupil

6 test tubes
Labels
Test tube rack

250 ml beaker
Tripod/wire gauze
Bunsen

Specimen result

	0°C	20°C	40°C	60°C	80°C	100°C	Ether	Conc Acid
2 min						▓	▓	▓
5 min					▓	▓	▓	▓
15 min				▓	▓	▓	▓	▓

Think about it!

1   The membrane is not permeable to the red substance. We know this because the substance normally stays inside the cells.
2   Heat affects the membrane so that it can no longer hold in the red substance.
3   Perhaps the heat caused the membranes to tear, etc, letting the red substance spill out.
4   Ether and acid damage the membrane. We know this because they caused the red substance to leak out.
5   Perhaps it is not the membrane which is holding in the colour but something else.

Suggested conclusion

The cell membrane is selectively permeable, allowing only certain substances through. Heat and certain chemicals damage it so that it becomes completely permeable.

7.7   Turgor and plasmolysis in plant cells

Object

To induce, observe and reverse plasmolysis in plant cells

Material per class

Onion separated into layers and cut into 1 cm squares or pieces of rhubarb
petiole (stalk)

Material per group/pupil

Microscope slide
Cover slip
Tissues
Filter paper triangles
Mounted needle

20% sucrose solution with dropper
Water with dropper
Scalpel
Forceps
Microscope with bench lamp if necessary

Specimen result

None

Think about it!

1   When bathed in water, water entered the plant cells since the water
     concentration inside the cells was lower.
2   Sucrose solution caused water to leave the cells since this has a lower water
     concentration than the contents of the cells.
3   With its cells plasmolysed, the plant would be wilted.
4   Plasmolysis is not a permanent state. It is reversed by adding water.
5   When a wilting plant is watered, the cells absorb water by osmosis. The
     cell contents press against the cell walls, making the cells turgid. This
     straightens out the leaves.
6   The cell wall does not keep out sucrose since otherwise the membrane and
     wall would stay together.

Suggested conclusion

When the bathing solution has a higher water concentration than the cells, water
enters by osmosis, making the cells turgid (stiff). This gives a plant support.
When the bathing solution has a lower water concentration, water leaves the cells,
the membrane comes away from the wall and the cells are plasmolysed. The plant
wilts. Plasmolysis is reversed if the plant is given water.

7.8 Osmosis and turgor in plant tissue

Object

To show how water loss or gain by osmosis can affect the size and turgor of plant tissue

Material per class

Large potatoes
Cork borer, 8 mm
Knives or scalpels

5% sucrose solution, 20 ml per group
20% sucrose solution, 20 ml per group

Material per group/pupil

3 test tubes
Test tube rack
Tissues

Ruler
Labels

Specimen result

Cylinder	Bathing solution	Length at start	Length at finish	Change in length +/-	Condition firm/normal/ flabby
A	Water	35 mm	40 mm	+5 mm	Firm
B	5% sucrose	35 mm	35 mm	0	Normal
C	20% sucrose	35 mm	30 mm	-5 mm	Flabby

Think about it!

1   5% sucrose solution has a higher water concentration than 20% sucrose solution.
2   If a potato cell took in water, it would get longer.
3   Cylinder A absorbed water.
4   Cylinder C lost water.
5   The water concentration of the cell contents would be about the same as the bathing solution.
6   The cells of cylinder C would be plasmolysed.

Suggested conclusion

If the bathing solution has a higher water concentration than the cell contents, water will enter the potato by osmosis, making the cylinder longer and firmer. If the bathing solution has a lower water concentration, water will leave the cells by osmosis, making the cylinder shorter and softer. If the water concentration is the same, there will be no effect.

7.9   Osmosis and red blood cells

Object

To observe haemolysis and crenation of red blood cells

Material per class

Sterile lancets (An alternative is the Autolet available from Philip Harris. This
   is a spring-loaded device which punctures the skin with a small, sterile lancet.)
Ethanol (industrial methylated spirit)
Cotton-wool, small wads in a clean dish
Elastic bands
Container for used lancets
0.85% salt solution in labelled dropping bottle
2% salt solution in labelled dropping bottle
Beaker of distilled water with dropper

Material per group/pupil

3 microscope slides   Mounted needle
Cover slip   Microscope, with bench lamp
Labels   if necessary

Comments

If pupils are to take their own blood, it is advisable that they be given some kind
of authorisation form for their parents to sign.
Great care must be taken to prevent infection.
It may be easier for the teacher to donate the blood.

Specimen result

None

Think about it!

1   The higher water concentration is outside the blood cells.
2   Osmosis is the diffusion of water across a semi-permeable membrane from high
     water concentration to low water concentration.
3   Water should move into the cells.
4   The cells eventually swell up and burst.
5   Plants' cell walls stop them bursting.
6   The cells in the 2% salt solution shrivel up.
7   The water concentration is higher inside the cells.
8   The water concentration of 0.85% salt solution must be the same as the cell
     contents because they are unchanged.
9   If a water concentration of the blood varied by too much, the blood cells would
     be damaged.

Suggested conclusion

The water concentration of red blood cells is about the same as 0.85% salt solution. In lower water concentrations (2% salt solution), they lose water by osmosis and shrink. In higher water concentrations (distilled water), they absorb water by osmosis, swell up and burst.

7.10 Water loss from plants

Method A: Source of loss

Object

To show that plants lose water largely from their leaves

Material per class

2 pot plants, eg geranium   Thread
2 polythene bags   2 stoppered bell jars

Specimen result

The inside of jar A becomes covered in condensation.

Think about it!

1   You could test its boiling and freezing points or use cobalt thiocyanate paper.
2   Plants lose water from their leaves.

Suggested conclusion

Plants lose a lot of water from their leaves. This is called transpiration.

Method B: Rate of loss

Object

To measure the transpiration rate of a leafy shoot

Material per group/pupil

Freshly-cut leafy shoot   Liquid paraffin with dropper
2 test tubes   Spirit marker
Test tube rack

Specimen result

Start time: 0900 Mon   Finish time: 1000 Tue   Time of experiment: 25 hrs

Loss from tube A: 1 ml   Loss from tube B: 0

Think about it!

1   Water evaporating would cool the leaves down.
2   This would be an advantage in summer and a disadvantage in winter.
3   Leaves have air pores called stomata which allow water molecules to escape
     from them.

Suggested conclusion

The leafy shoot lost water by transpiration at the rate of ... ml/hour.

Method C: Loss from upper and lower epidermis

Object

To show that most of a leaf's water loss is through the lower epidermis

Material per class

Balance, as accurate as possible   Paper clips
Dicotyledon, eg privet, leaves   Thread
Vaseline

Specimen result

Leaf	Surface covered	Weight at start (m)	Weight at finish (n)	Weight loss (m-n)	Percentage weight loss $\frac{m-n}{m}$ x 100
A	Upper	0.9 g	0.5 g	0.4 g	44%
B	Lower	0.7 g	0.6 g	0.1 g	14%

Think about it!

1　　If the leaves lose water, they will lose weight.
2　　Vaseline could clog up the stomata (air pores).
3　　The lower surface allowed more water to escape through it.
4　　The lower surface has more stomata.
5　　The upper epidermis has few stomata and a waxy cuticle to reduce water loss.
6　　This is not completely efficient since leaf B did lose some weight.
7　　Since the leaves are not exactly the same size, this gives a better picture of what is happening.

Suggested conclusion

Most of a leaf's water loss is through the many stomata on the lower epidermis. The cuticle on the upper epidermis allows a little water through.

7.11 Root hairs

Object

To observe the location, structure and water absorption of root hair cells

Material per class

Cress seedlings growing on discs of black crepe paper about 1 cm across. They are
　germinated in a covered Petri dish in a warm place with the discs being
　moistened regularly.
Forceps
Hand lenses
Binocular microscopes
Iodine solution in labelled dropping bottles (see page 6)
Beaker of dye, eg eosin, red ink or safranin with dropper

Material per group/pupil

2 microscope slides
Cover slips
Scalpel

Mounted needle
Microscope, with bench lamp
　if necessary

Specimen result

(i)　Drawing to show the position and size of the root hairs on the cress seedling
(ii)　Drawing of single root hair cell labelled as in Step 5
(iii)　Drawings to show colour penetrating root hair before the rest of the root

Think about it!

1   The youngest root hairs are those nearest the tip.
2   Root hair cells are only found in one part of the root.
3   The nucleus of a root hair cell is at the end.
4   Water enters the root by osmosis into the root hairs.
5   The large surface area of the root hairs makes water absorption by the root more efficient.

Suggested conclusion

Root hairs increase a root's surface area so that it can absorb more water. Water is passed, by osmosis, from the root hairs into the root itself.

8.1   Cyclosis/Cytoplasmic streaming

Object

To observe cytoplasmic streaming in plant and animal cells

Material per class

Paramecium caudatum, obtained from biological suppliers. It can be cultured as
  outlined on page 7.
Yeast cells stained with congo red as follows. Add 0.5 g of congo red to 250 ml of
  water in a large beaker. Add 5 g of finely powdered yeast and boil gently for
  15 minutes. Allow to cool, provide a dropper.
Polycell solution, equal parts Polycell and water, with a dropper
Elodea leaves from the top of actively-growing shoots, in water
Beaker of water with dropper

Material per group/pupil

2 microscope slides   Microscope, with bench lamp
Cover slips   if necessary
Mounted needle

Specimen result

None

Method A: Think about it!

1   The cilia look like a shimmering border.
2   The cilia are drawing in the yeast cells.
3   The Paramecium is digesting the yeast since the dye inside it is changing
     from red to blue.
4   The yeast takes a figure-of-eight path inside the cell.

5   Moving its food around would help to distribute it round the cell.

Method B: Think about it!

1   The cytoplasm could be moving and carrying the chloroplasts around with it.
2   ...
3   Such movement might help to distribute material around the cell.

Suggested conclusion

In Paramecium, the cytoplasm moves round in a figure-of-eight path, digesting food as it goes. In Elodea, it goes round in a circle, taking chloroplasts with it. These movements may help to distribute things round the cell.

8.2   Looking at blood

Object

To prepare and observe a blood smear

Material per class

Sterile lancets (An alternative is the Autolet available from Philip Harris. This is a spring-loaded device which punctures the skin with a small, sterile lancet.)
Ethanol (industrial methylated spirit)
Cotton-wool, small wads in a clean dish
Elastic bands
Container for used lancets
Phosphate buffer pH 6.8 from biological supplier, with dropper
Leishman's stain in labelled dropping bottle
Prepared slides

Material per group/pupil

2 very clean microscope slides
Microscope, with bench lamp if necessary

Comments

If pupils are to take their own blood, it is advisable that they be given some kind of authorisation form for their parents to sign.
Great care must be taken to prevent infection.
It may be easier for the teacher to donate the blood.

Specimen result

None

Think about it!

1   Red blood cells are most abundant.
2   Red blood cells have no nucleus.
3   The most outstanding part of white blood cells is their nucleus.

Suggested conclusion

Blood contains various kinds of cell. Most are red blood cells which have no nucleus. There are several kinds of white blood cells which have prominent nuclei of different types.

8.3   Pulse rates

Object

The effect of exercise on the pulse rate

Material per class

None

Specimen result

None

Think about it!

1   The pulse rate is a measure of the heart's rate of beating.
2   Exercise increases the pulse rate.
3   During exercise, the muscles must have more oxygen. Breathing and heart rate increase to supply this and also to get rid of the extra carbon dioxide produced.
4   During exercise, lactic acid, produced by anaerobic respiration, builds up in the muscles. After extra exercise, extra oxygen is needed to get rid of this.
5   Fit people do recover faster.

Suggested conclusion

The pulse measures how fast the heart beats. During exercise, the heart has to beat faster to supply more oxygen to the muscles. After exercise, it must continue to work hard until lactic acid is removed - until the 'oxygen debt' is paid off. Fit people recover faster.

8.4 Water transport in plants

Object

To show the presence and distribution of water-carrying tissue in plants

Material per class

Leafy plants, eg leek, celery, busy Lizzie, groundsel, wandering sailor and most
    soft green annual plants
Water-soluble dye, eg eosin, safranin and red ink
Carrots or elder pith
Water with dropper

Material per group/pupil

Scalpel or razor blade
Hand lens
Beaker of water with dropper
Microscope slide
Cover slip

Mounted needle
Watch glass
Microscope, with bench
    lamp if necessary

Specimen result

None

Method A: Pathways

Think about it!

1   We used coloured water so that we could follow its progress.
2   Plants draw up water only through certain parts of their stems.
3   Yes. The colour stays in lines running up the plant.
4   You could time when the plant was put in and taken out of the dye and measure
     how far up the plant it had gone in that time.

Suggested conclusion

Water travels up a stem through a system of veins.

Method B: Veins

Think about it!

1   The plant did manage to raise water up to its leaves since the colour reached
     them.
2   Water only travels through certain parts of the leaf.
3   The leaf's water supply system is a series of veins running through.

Suggested conclusion

Water is transported through leaves in a system of veins.

Method C: Transport tissues

Think about it!

1   No. Water goes up the middle of a root, but it goes up veins around the outside of a stem.
2   The water-carrying cells look round with thick walls.
3   Water-carrying cells in the root are similar to those in the stems.
4   No other contents were visible in the water-carrying cells.

Suggested conclusion

The veins in a root are in the centre, while those in a stem are round the outside. The water-carrying vein cells are circular in transverse section, empty-looking with thick walls.

9.1   Surface area to weight ratios

Object

To show that small organisms have a larger surface area to weight ratio than large ones and that shape makes a difference

Material per class

Centimetre tape measures (or string and rulers)
Bathroom scales
Balance, Butchart or triple beam

Specimen result

None

Think about it!

1   A large organism has more surface area than a small one.
2   A large organism has more weight than a small one.
3   A small organism has a higher surface area to weight ratio than a large one.
4   A large organism will have more cells than a small one.
5   A small organism will have more surface for each cell than a large one.
6   A small organism will have more skin in proportion to its weight than a large one.
7   The shape of an organism will affect its surface area to weight ratio. A long, thin organism has a higher surface area to weight ratio than a short, fat organism of the same weight.

Suggested conclusion

Large organisms have a smaller surface area to weight ratio than small organisms. Long, thin organisms have a higher ratio than short, fat organisms of the same weight.

9.2 Surface area to weight ratio and heat loss

Object

To show that a small body (large SA/W) loses heat more quickly than a large body (small SA/W)

Material per group/pupil

100 ml round-bottomed flask fitted with stopper and thermometer
500 ml round-bottomed flask fitted with stopper and thermometer
2 retort stands with clamps
Stopclock

Specimen result

Flask	Surface area	Weight of water	Surface area / Weight
Small	115 cm²	100 g	1.15
Large	330 cm²	500 g	0.66

Time (min)	0	5	10	15	20	25	30	35	40	45	50	55	60
Small flask temperature (°C)	85	80	73	67	63	59	55	53	50	47	45	43	41
Large flask temperature (°C)	85	82	78	74	71	68	65	62	60	58	55	53	51

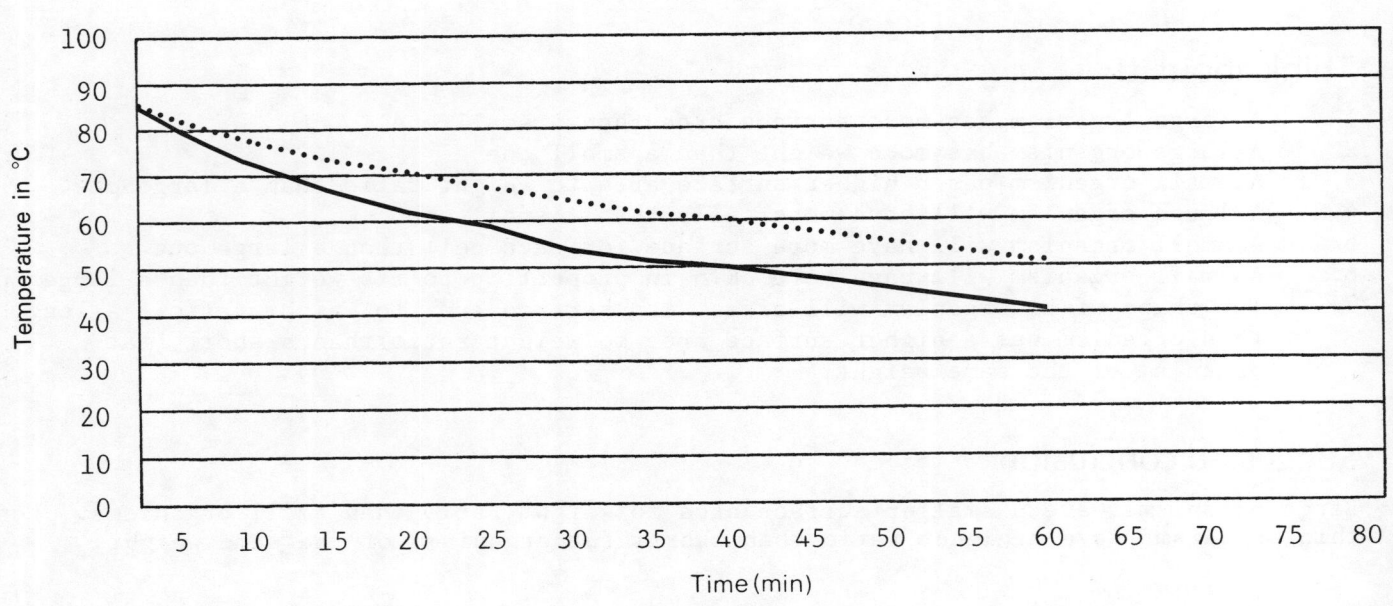

Think about it!

1   The flask with the larger surface area to weight ratio cools down faster.
2   Birds and mammals are warm-blooded and large ones will cool down less in cold climates than small ones would.
3   Small birds and mammals lose heat more quickly than large ones and so are better suited to hot regions.
4   Dogs lose heat by panting. Indoors, a small dog will lose enough heat through its skin to be comfortable.
5   The large flask contained more energy at the start.
6   The large flask had most surface in contact with the air.
7   The large flask would lose more heat energy.
8   A large animal contains more heat than a small one and so can afford to lose more without cooling down as much.

Suggested conclusion

A small body has a larger surface area to weight ratio than a large one and so cools down more quickly. This means that small warm-blooded animals are more suited to warm climates than large ones.

9.3   Surface area and water loss

Object

To show the effect of surface area to weight ratios on water loss from a body

Material per group/pupil

100 ml beaker
2 250 ml beakers
Measuring cylinder, 200 ml

Specimen result

Beaker	Weight of water at start k	Weight of water at finish l	Weight of water lost (k-l)	Percentage water lost $\frac{(k-1)}{k} \times 100$	Surface area (cm²)	Surface area / Weight
A	25 g	19 g	6 g	24	20	0.8
B	25 g	15 g	10 g	40	37	1.48
C	150 g	14 g	9 g	6	37	0.25

Comments

The surface area of the liquid is $\frac{\pi}{4} d^2$. This is approximately $0.8 d^2$.

Think about it!

1   More weight of water is lost through a large surface area since beakers B and C lost more than beaker A.
2   A large body has more surface area than a small body.
3   A large organism would lose more water through its skin than a small organism.
4   A large organism could tolerate water loss more than a small organism since beaker C lost a smaller percentage of its water than beaker B.
5   A small organism has a higher surface area to weight ratio than a large one.
6   The shape of an organism does affect water loss through its skin since beaker B lost more water than beaker A, though they both started with the same weight.

Suggested conclusion

The larger the surface area, the greater will be the water loss. However, a large organism will have a smaller surface area to weight ratio than a small one and so should be able to tolerate water loss better. Also, a thin, flat organism will lose water more quickly than a thick, compact one of the same weight.

9.4   Stability and weight

Object

To show the effect of length and disposition of legs on an animal's stability and body weight

Material per group/pupil

100 g lump of Plasticine
Drinking straws or wooden
    dowelling

Small platform, wood or card, with
    raised rim along one edge
Protractor

Specimen result

Animal	Angle of tilt	Weight supported
Short, straight legs	15°	100 g
Short, splayed legs	27°	60 g
Long, straight legs	8°	80 g
Long, splayed legs	19°	40 g

Think about it!

1   The animal with the short, splayed legs has the lowest centre of gravity.
2   The animals with splayed legs have the biggest base.

3   The animal with short, splayed legs is most stable.
4   Legs pointing straight down can support most weight.

Suggested conclusion

A low centre of gravity and wide base gives greatest stability. Splayed legs can
support less weight than legs directed straight down under the body.

9.5   The strength of bones

Object

To show that short bones will be stronger than long bones and that hollow bones would
be stronger than solid bones of the same weight

Material per class

Perspex safety screen
Glass rods
Glass tubing of the same weight per
    centimetre as the glass rod

Scale pan with attached string
100 g weights
2 retort stands with clamps

Specimen result

Glass tested	Length	Breaking strength
Solid	32 cm	400 g
Solid	16 cm	700 g
Solid	8 cm	1000 g
Solid	4 cm	1300 g
Hollow	32 cm	600 g

Think about it!

1   Legs are long bones and so are more easily broken than short finger bones.
2   Shorter leg bones will support a heavier body than long bones.
3   A hollow bone is stronger than a solid bone of the same weight.

Suggested conclusion

Short bones are stronger than long bones. Hollow bones are stronger than solid
bones of the same weight.

9.6   The components of bone

Object

To show the different qualities of the organic and mineral components of bone

Material per group/pupil

3 small, similar chicken bones. The wing bones
   from packs of 'chicken bites' are ideal
2 boiling tubes
Test tube rack
Labels

Dilute hydrochloric acid
Forceps
Tripods
Pipe clay triangle
Bunsen

Specimen result

	Treatment	Component removed	Appearance of bone	Condition of bone
Bone A	Water	None	Normal	Normal
Bone B	Acid	Mineral	Normal	Bendy
Bone C	Roasting	Organic	Charred	Easily broken

Think about it!

1   A bone should be slightly flexible to prevent it breaking easily under strain.
2   The protein component gives bone its flexibility.
3   It is too soft. The bone bends too easily.
4   The mineral component of bone gives it its hardness.
5   On its own, the mineral component is brittle and easily broken.

Suggested conclusion

Bone consists of an organic component (protein fibres) which gives it flexibility
and a mineral component (calcium phosphate) which gives it hardness.

9.7   Muscle force

Object

To show the effect of fatigue on muscles and that regular use increases the force a
muscle can apply

Material per class

Bathroom scales marked in Newtons or kilograms (multiply kilograms by 10 to convert
 to Newtons)

Specimen result

None

Think about it!

1   I can feel the arm muscles hardening as I squeeze.
2   The muscles are connected to the fingers by tendons.
3   A muscle contracts (gets shorter) to move a joint.
4   It must use up energy to contract.
5   Repeated contraction causes a muscle to get weaker. The muscle could have used
        up its energy stores.
6   Rest allows the muscle to recover its force. The muscle could have replaced
        its energy supplies.
7   Training increases the force a muscle can apply and allows it to work longer.

Suggested conclusion

Repeated contraction of a muscle causes tiredness or fatigue since it uses up the
muscle's energy supplies. Rest allows the muscle to replace these so it recovers.
Regular training increases the force a muscle can apply and allows it to work longer.

10.1 Skin sensitivity

Object

To show the presence and examine the distribution of sense receptors in the skin and
to investigate its perception of temperature

Material per group/pupil

1 pair of dividers, preferably with adjusting screw
Ruler
3 beakers, eg 250 ml
6 long, thin nails
Crushed ice
Tongs
Rubber pad with a grid of 36 raised dots. This is made by carefully drawing a grid
 of fine lines, 1 mm apart, on an ink eraser. A sharp scalpel or razor is then
 used to cut 1 mm wide grooves to form a square, 6 dots by 6 dots. A felt pen can
 be used to ink it.

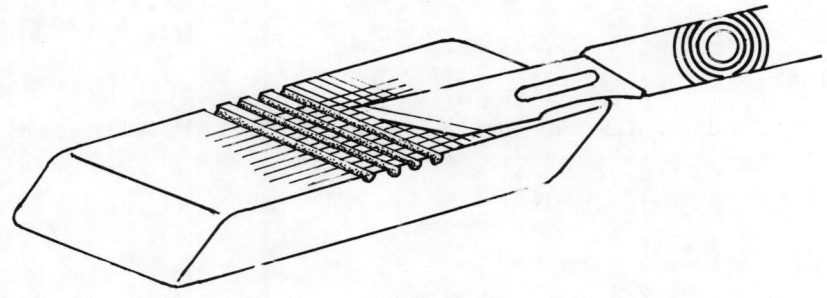

Method A: Two-point thresholds

Specimen result

The two-point thresholds vary between individuals. Small values should be obtained for the fingertips, palm and sole of the foot. Larger ones are found for the forearm, back of the neck and cheek.

Think about it!

1   If the two points are very close together, they will only touch one receptor.
2   Touch receptors in the skin are closest together in the fingertips, palm and sole. They are further apart in the forearm, back of the neck and cheek.

Suggested conclusion

Touch is detected by receptors under the skin. These are very close together in the fingertips, palm and sole but further apart elsewhere.

Method B: Temperature receptors

Specimen result

The 'heated' finger felt cold in the lukewarm water while the 'cooled finger' felt warm. There are more 'cold spots' in the skin than 'hot spots' and the distribution of both varies as for touch.

Think about it!

1   The lukewarm water felt cold to the 'heated' finger and hot to the 'cooled' finger.
2   No. Both fingers were actually at the same temperature.
3   The 'heated' finger was losing heat in the lukewarm water. It felt cold.
4   The 'cooled' finger was gaining heat in the lukewarm water. It felt hot.
5   The skin actually measures heat loss or heat gain and not temperature.
6   Only certain spots on the skin respond to heat loss or gain.
7   A spot on the skin responds either to heat loss or heat gain but not both.
8   Heat and cold receptors are thought to be under the skin at 'hot spots' and 'cold spots'.

Suggested conclusion

There are separate heat and cold receptors in the skin. These respond to heat loss or gain from the skin and not to temperature.

10.2   Forming an image

Object

To show that convex lenses can form inverted colour images on a screen and that thick lenses give clear images of near objects while thin lenses give clear images of distant objects.

Material per group/pupil

2 convex lenses - a 100 mm focus and a
  50 mm are suitable
Lensholder
Screen (if necessary, a sheet of white
  card or paper is suitable)

Metre stick
Candle or other suitable light
  source, preferably with an
  obvious 'top' and 'bottom'

Specimen result

Distance between object and thin (100 mm) lens = 88.5 cm
Distance between object and thick (50 mm) lens = 79.5 cm

Think about it!

1   The image is inverted.
2   The image is in colour.
3   A thin lens gives a clear image of a distant object.
4   A thick lens gives a clear image of a near object.
5   The image gets smaller as the object gets farther away.

Suggested conclusion

Convex lenses form inverted images of an object in colour.
A thin lens forms a clear image of a distant object.
A thick lens forms a clear image of a near object.

10.3   How lenses work

Object

To show how various prisms and lenses affect the path of light rays passing through them

Material per group/pupil

Raybox with slit plate
Various glass or plastic blocks as shown in <u>Experiment Guide</u>. (These can be
    obtained separately or in kit form from Griffin and Philip Harris.)

Specimen result

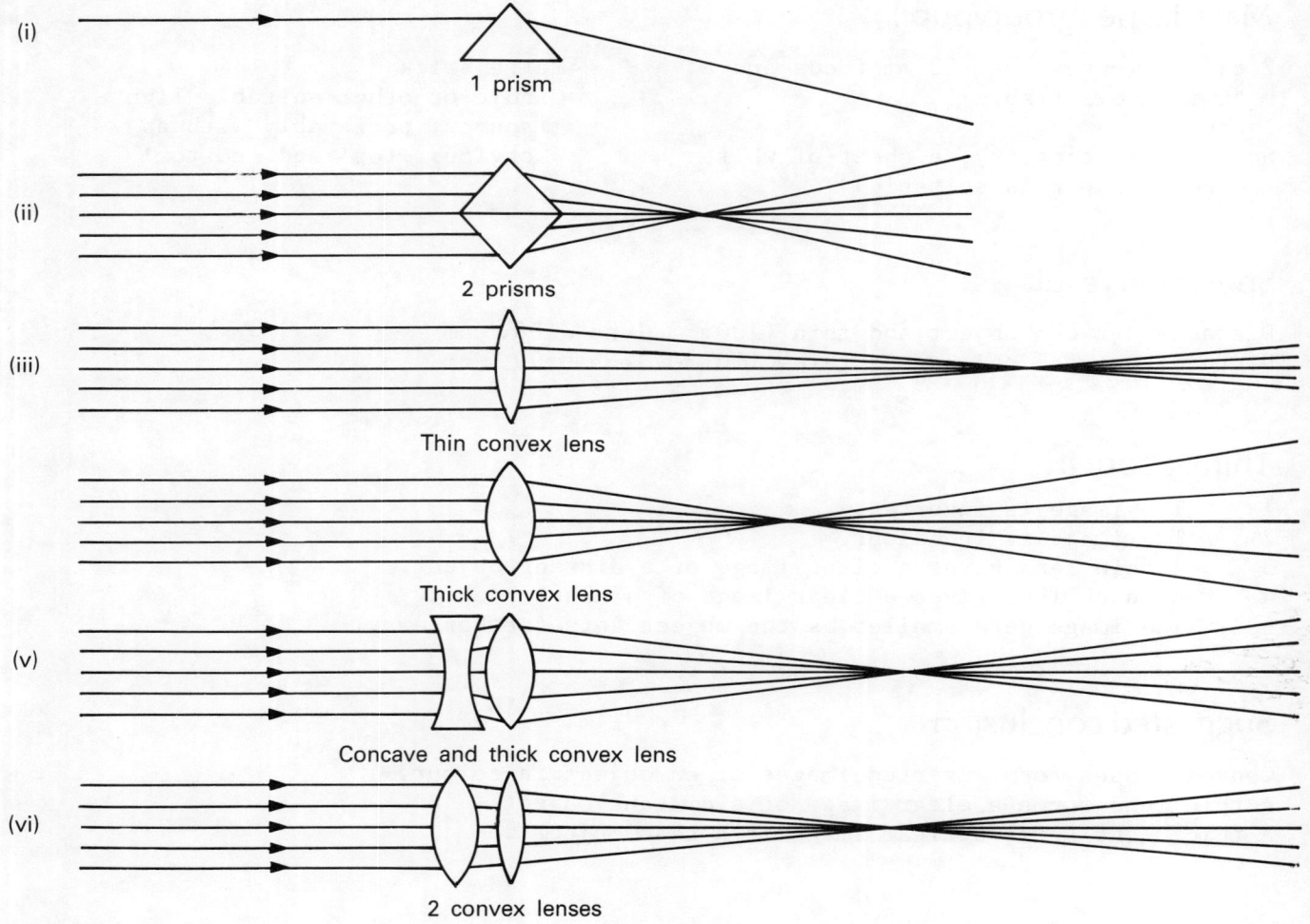

Think about it!

1   A prism bends light rays.
2   Two prisms will bring a row of parallel light rays to a point (or focus).
3   The two prisms resemble a convex lens.

4   Lenses work by bending light rays.
5   The eye has a convex lens.
6   This lens bends them inwards so that they meet.
7   A thick lens bends light more than a thin lens.
8   If the lens in a eye was bending light too much, a concave lens would correct this.
9   If the lens in an eye was not bending light enough, a convex lens would correct this.

Suggested conclusion

Lenses work by bending rays of light. Convex lenses bend them inward so that they meet at a point called the focus. Thick lenses bend them more than thin. Concave spectacle lenses will correct overbending by the lens in the eye (short sight) and convex lenses will correct underbending (long sight).

10.4   Some limits of vision

Object

To demonstrate the existence of the blind spot, differences in colour sensitivity, and the use of binocular vision

Material per class

Sets of felt pens, or colour pencils, clearly marked with their colours on case or lid. Strip of paper fixed to the wall around eye height when seated. To its left is marked a bold spot, then to the right of this, a range of angles at the following distances from the cross: 10°-6 mm; 15°-8 mm; 20°-11 mm; 25°-14 mm; 30°-18 mm; 35°-22 mm; 40°-26 mm; 45°-31 mm; 50°-37 mm; 55°-44 mm; 60°-53 mm; 65°-65 mm; 70°-83 mm; 75°-112 mm; 80°-166 mm.

Specimen result

None

Method A: Blind spot

Think about it!

1   The spot disappears when its image falls on the blind spot of the retina.
2   In its place is its background.
3   The blind spot is not big since a slight movement of the eye is enough for its effect to disappear.

Suggested conclusion

The blind spot on the retina is where the optic nerve leaves the eye. At that point, there are no light receptors. The brain fills in the gap in vision so that we are not aware of it.

Method B: Colour sensitivity

Think about it!

1   Cones detect colours.
2   Rods detect dim light but not colour.
3   No. There are not colour receptors all over the retina since it is possible to see the pens but not know their colour.
4   There is more than one kind of colour receptor since all colours are not detected equally well.
5   The retina is most sensitive to ... and least sensitive to ...
6   ... might be the best colour for an emergency vehicle.

Suggested conclusion

Parts of the retina contain rods but no cones and so cannot detect colour. The different types of cones are not present in equal numbers in the retina. The retina is most sensitive to ... and least sensitive to ...

Method C: Binocular vision

Think about it!

1   The two eyes are not in exactly the same position and so each gets a slightly different picture of an object.
2   We are not usually aware of this since the brain merges the two pictures into one.
3   This can be useful in judging distances.

Suggested conclusion

Each eye sends a slightly different picture to the brain. This is called binocular vision and allows us to judge distances more accurately.

10.5   Reaction times

Object

To measure the visual and auditory reaction times

Material per group/pupil

Commercial reaction timing apparatus, if available, or strips of card marked in centimetres and seconds as shown in the Experiment Guide (a master could be photocopied then stuck to strips of stiff card), or a metre or half metre stick.

Specimen result

None

Think about it!

1   Practice does improve performance.
2   The delay is caused by the time taken for signals to travel along nerves and across synapses.
3   Synapses are gaps between adjoining nerve cells.
4   More synapses would increase the reaction time.
5   ....
6   ....

Suggested conclusion

The reaction time is the time taken for signals to travel along nerves and across synapses. Practice can shorten reaction times to some extent. The auditory reaction time is longer than the visual since there are more synapses in the auditory pathways.

10.6   Reflexes

Object

To demonstrate the existence of the knee-jerk and pupillary reflexes

Material per group/pupil

Metal rod, eg a retort stand rod
Torchlight

Specimen result

None

Think about it!

1   The knee jerk was not a voluntary response.
2   The response could not really be stopped.

3   The pupil of the eye is a hole in the iris.
4   Bright light causes the pupil to become smaller. Dim light causes it to become bigger.
5   The pupillar reflex will protect the eyes from bright light.
6   No. Both eyes respond together.

Suggested conclusion

A tap on the knee causes the leg to jerk out. This is an involuntary response. Pupils widen in dim light and become smaller in bright light. This protects the eyes. Both pupils respond together.

10.7   Responding to the environment

Object

To show the effect of environment on the behaviour of woodlice and <u>Paramecium</u>

Method A: Woodlice and light

Material per group/pupil

A choice chamber made from Petri dishes as shown in the <u>Experiment Guide</u>. The dishes can be cut using a fine hacksaw, then rough edges filed down. The connecting tunnel is made from card or plastic glued in place with plastic adhesive. The whole assembly should be glued to stiff card or wood.
Blotting paper and scissors
Water, with dropper
Bench lamp
Black paper box to fit over chamber as shown in the <u>Experiment Guide</u>
12 woodlice of same species. Woodlice can be kept until needed in a moist habitat containing rotten wood, leaf litter, etc, and some pieces of raw potato.

Method B: *Paramecium* and chemicals

Material per group/pupil

Microscope slide
Cover slip
Mounted needle
Microscope, with bench lamp if necessary
Weak sodium bicarbonate solution with fine dropper

<u>Paramecium caudatum</u> obtained from a biological supplier and cultured as outlined on page 7.
Very weak acetic acid solution with fine dropper

Methods A and B: Specimen result

None

Methods A and B: Think about it!

1   They move about more until they are out of the light, then they slow down.
2   This response is called phototaxis.
3   In this case, the response is negative.
4   The woodlice's response to light is called negative phototaxis.
5   Sodium bicarbonate solution causes positive chemotaxis in Paramecium.
6   Acetic acid causes negative chemotaxis in Paramecium.
7   The first response would draw it towards food. The second response would
     send it away from harmful conditions.

Method A: Suggested conclusion

Woodlice seem to 'prefer' dark conditions and so show negative phototaxis.

Method B: Suggested conclusion

Paramecium shows positive chemotaxis for weak acids which indicate food but
negative chemotaxis for stronger acids which could be harmful to it.

10.8   Positive geotropism in roots

Object

To show positive geotropism in roots and demonstrate that the response is caused
by unidirectional gravity

Material per class

Clinostats. While each group should be able to set up a seedling with its radicle
   horizontal, it may be more convenient to demonstrate the effect of a moving
   clinostat.

Material per group/pupil

2 broad bean or pea seedlings with radicles about 2 cm long. These are prepared as
   outlined in the Experiment Guide, Section 10.
2 large Petri dishes
Cotton-wool strips about 30 mm wide, the full thickness of the roll
Elastic bands
Dropper or dropping bottle of water
Label or marker

Specimen result

None

Think about it!

1   Radicle A grew downwards.
2   Gravity seems to cause this type of growth.
3   A root will not bend if gravity affects it equally from all directions.
4   Gravity at one side of a root will cause it to bend downwards towards that side.
5   This response is called positive geotropism.
6   Positive geotropism is of use to a plant since it makes its roots grow down into the soil. Also no matter which way up a seed lands on the ground, its roots always grow downwards.

Suggested conclusion

If a root is subjected to gravity from one side only, it will bend towards that side. This is called positive geotropism.

10.9   The sensitive part of a root

Object

To show that the response to gravity made by a root takes place near the tip

Material per class

Container for Petri dishes, perhaps a sandwich box with Plasticine to hold the dishes in position.

Material per group/pupil

2 broad bean or pea seedlings with radicles about 2 cm long. These are prepared as outlined in the Experiment Guide, Section 10.
Soft tissues
Watch glass with Indian ink
Blotting paper
Hair grip or paper clip with thread stretched across it

Large Petri dish
Cotton-wool strips about 30 mm wide, the full thickness of the roll
Elastic band
Dropper or dropping bottle of water
Label or marker
Scalpel

Comments

After marking, a result should be obtained within 3 or 4 days.

Specimen result

None

Think about it!

1   Both radicles were exposed to gravity from the same direction.
2   They did not respond in the same way. The cut one did not grow. The other grew downwards.
3   We call the response of the intact radicle positive geotropism.
4   Gravity from one side caused the response.
5   The tip of the radicle seems to be responsible for detecting and responding to the stimulus.
6   The ink marks are not all the same distance apart.
7   The ink marks near the tip are farther apart.
8   The part of the root just behind the tip grows most since the marks are further apart there.
9   The part of the root which bends is also the part which is growing.

Suggested conclusion

The last few mm of a root seem to be able to detect and respond to gravity. Growth may be involved in this since this is also the part of the root where most growth takes place.

10.10 The effect of one-sided light on stems

Object

To show positive phototropism in plants and demonstrate that the shoot tip detects and responds to the unidirectional light

Material per class

Cardboard boxes with slit cut in one end as shown in the Experiment Guide, Section 10.
Bench lamps
Cooking foil
Scissors
Matchsticks

Material per group/pupil

3 small Petri dishes, each containing 6 to 12 seedlings of oat or other cereal about 20 mm long growing on cotton-wool. (About a week before needed, soak the grains for 24 hours. Plant them, right way up, in the Petri dishes lined with a half thickness of cotton-wool. Keep the cotton-wool moist. When the seedlings are

the desired height, keep them in the fridge until required.)
Labels or marker
Dropper or dropping bottle of water

Specimen result

None

Think about it!

1 One-sided light causes plants to bend towards it.
2 The response of B is called positive phototropism.
3 The stimulus causing the response is one-sided light.
4 The foil caps of C prevent the response.
5 The tip of the plant stem seems to detect and respond to one-sided light.

Suggested conclusion

When a plant receives light from one side only, the stem bends towards it. This is called positive phototropism. The tip of the shoot seems to detect and respond to the stimulus.

10.11 The effect of indole acetic acid (IAA) on stems

Object

To show the effect of indole acetic acid on the growth of plant stems

Material per class

0.1% IAA in lanolin
Pure lanolin
  These can be obtained from biological suppliers or prepared by the method outlined
  on page 7.   This must be kept warm for ease of application. Small amounts could
  be dispensed in labelled watch glasses or test tubes when required. Surplus should
  be discarded.
Matchsticks, toothpicks, etc for application

Material per group/pupil

4 small Petri dishes, each containing 6 to 12 seedlings of oat or other cereals
  about 20 mm long prepared as on page 8.
Labels or marker
Dividers
Ruler
Dropper or dropping bottle of water

Comments

It may be more convenient and economical to let each group do either A,B,C or D and combine results.

Specimen result

	Average length at start (m)	Average length at finish (N)	Average change in length (n-m)	Percentage change $\frac{n-m}{m} \times 100$
Group A	20 mm	28 mm	8 mm	40%
Group B	20 mm	38 mm	18 mm	90%

Think about it!

1   If IAA is put on the tip of a shoot, the shoot will increase greatly in length.
2   If IAA is put on the side of a shoot, the shoot bends towards the other side.
3   If it makes one side grow more than the other, then the shoot will bend to the slow-growing side.
4   It could not have been the lanolin since dish D showed much less growth than dish B.
5   Lanolin on dish A would have no effect.
6   The coleoptiles were kept in the dark so that they could not be affected by one-sided light.

Suggested conclusion

Indole acetic acid causes increased growth in shoots. If one side of a shoot gets more IAA than the other

11.1   Reproduction in yeast

Object

To observe budding in yeast and to show that this leads to an increase in numbers.

Material per class

5% glucose solution, 5 ml per group
Yeast suspension. Add 5 g of dried yeast to 250 ml of warm (30°C) 5% glucose solution.
  Keep in a warm place for 2 days. Stir gently before use.
5 ml syringes
1 ml syringes
Test tube racks

Material per group/pupil

Test tube
Label
Glass rod
Microscope slide

Cover slip
Mounted needle
Microscope, with bench lamp
   if necessary

Specimen result

None

Think about it!

1   Yeast is a single-celled fungus.
2   The culture looked 'alive' since it was bubbling.
3   The yeast feeds on the glucose in this experiment.
4   The yeast is reproducing since its numbers are increasing.
5   The yeast is reproducing by budding.

Suggested conclusion

Yeast is a single-celled fungus. Under suitable conditions, it reproduces rapidly
by budding.

11.2   Vegetative reproduction in angiosperms

Object

To dissect various organs of perennation of angiosperms and investigate the nature
of their food supplies

Material per class

Potato tubers, various corms and bulbs
Scalpel or knives
Hand lenses

Iodine solution (see page 6)
   in labelled dropping bottles
Benedict's solution (see page 6)
   in labelled dropping bottles

Material per group/pupil

Large beaker (400 ml)
Spotting tile
Test tube
Label

Tripod/wire gauze
Bunsen (or thermostatic water
   bath set at 80-90°C)

Specimen result

None

Think about it!

1   These organs are swollen with stored food.
2   They need a food store to provide for next season's growth.

Suggested conclusion

Tubers and corms are stems swollen with food for the next year's growth. Bulbs are swollen leaves.

11.3   Flower structure

Object

To dissect, examine and compare an insect- and a wind-pollinated flower

Material per class

Various insect- and wind-pollinated flowers. Large simple flowers such as spring 'bulbs' are suitable. Composites such as daisy and dandelion are awkward. Ripe grasses may be used for the wind-pollinated flowers. The insect- and wind-pollinated flowers can be examined at different times of the year - whenever they are available.
Binocular microscopes

Material per group/pupil

Hand lens	Microscope slide
Forceps	Microscope, with bench lamp
Scalpel	if necessary

Specimen result

None

Think about it!

1   Petals are only needed to attract insects.
2   The carpel does stick up in the insect-pollinated flower to make sure it touches the insects.

3   The carpel does not stick up in the wind-pollinated flower. It hangs outside
    to catch pollen on the wind.
4   The stamens do stand up in the insect-pollinated flower to make sure they touch
    the insects.
5   The stamens do not stand up in the wind-pollinated flower. They hang outside
    to scatter their pollen in the wind.
6   The wind-pollinated flower had more pollen since more of it gets wasted.
7   The wind-pollinated flower had smaller grains so that they could be carried
    by the wind.

Suggested conclusion

There are a number of differences between insect- and wind-pollinated flowers. Insect-
pollinated flowers have petals to attract insects, upright stamens and carpels to
touch the insects and large, sticky pollen. Wind-pollinated flowers have no petals.
Their stamens and carpels hang down to spread and catch pollen which is light and
produced in large quantity.

11.4   Growing pollen tubes

Object

To observe the growth of pollen tubes

Material per class

10% sucrose solution with droppers (see Comments below)
Incubator set at 20°C
Flowers with ripe pollen

Material per group/pupil

Large Petri dish, 90 mm   Filter or blotting paper
Microscope slide   Label
Forceps   Water and dropper
Visking, 4 cm x 2 cm   Microscope, with bench
Matchsticks   lamp if necessary
                                   Cover slip

Comments

The concentration of sucrose solution required varies with plant species. For
example, 10% is suitable for sweet pea, hyacinth, fuchsia and bluebell, 7% for
narcissus, 5% for apple and primrose and 3% for tulip.

Specimen result

None

Think about it!

1   The pollen usually sprouts a tube from the stigma of a flower.
2   The sugar solution produced by the stigma causes pollen to grow a pollen tube.
3   The pollen tube grows from the stigma, down the style and into an ovary. It does this so that it can penetrate an ovule and fertilise the female egg nucleus.
4   This would be an advantage because pollen grains would not germinate on the stigma of the wrong species.

Suggested conclusion

Sugar solution causes pollen grains to germinate and grow pollen tubes. In nature, the stigma of flowers produces this so that pollen tubes can grow down the style, into the ovary to fertilise the ovule.

11.5   The structure of seeds

Object

To examine the structure of a dicotyledon seed and a monocotyledon fruit and locate their food stores.

Material per class

Broad beans and maize seeds soaked in water for 48 hours
Iodine solution in labelled dropping bottles (see page 6)
Scalpels
Hand lenses

Specimen result

None

Think about it!

1   The broad bean has two cotyledons.
2   The cotyledons store the food in the broad bean seed.
3   The broad bean stores starch.
4   The broad bean's plumule is tucked between its cotyledons for protection.
5   The radicle is near the outside since this is the first part to grow.
6   Maize has only one cotyledon.
7   The endosperm stores the food in maize.
8   The maize fruit stores starch.
9   The pericarp is the part of the fruit which develops from the ovary wall.

Suggested conclusion

The two cotyledons store the food in the broad bean, while the endosperm stores it in maize. In both cases, the stored food is starch.

11.6 Conditions for seed germination

Object

To investigate the conditions required for germination of cress seeds

Material per class

Beaker of water which has been boiled
    then allowed to cool
Cress seeds

Liquid paraffin with dropper
Test tube racks (for fridge and dark)
Cotton-wool

Material per group/pupil

5 test tubes
5 labels

Test tube rack
5 stoppers or squares of Parafilm

Specimen result

Test tube	Did they have water?	Did they have oxygen?	Did they have warmth?	Did they have light?	Did they germinate?
A	Yes	Yes	Yes	Yes	Yes
B	No	Yes	Yes	Yes	No
C	Yes	No	Yes	Yes	No
D	Yes	Yes	No	Yes	No
E	Yes	Yes	Yes	No	Yes

Think about it!

1   Test tube B shows that seeds need water to germinate.
2   Test tube C shows that seeds need oxygen to germinate.
3   Test tube D shows that seeds need warmth to germinate.
4   Test tube E shows that seeds do not need light to germinate.
5   Before they germinate, seeds need water, oxygen and warmth but not light.

Suggested conclusion

To germinate, seeds need water, oxygen and warmth. They do not seem to need light.

11.7 *Drosophila:* the monohybrid cross

Object

To perform a monohybrid cross with Drosophila melanogaster to the F_2 generation

Material per class

Large stoppered jar containing 70% alcohol
Incubator set at 25°C
Ether in dropping bottles

Material per group/pupil

1 culture tube containing 6 male Drosophila, either wild type (tube A) or mutant
  (tube D), clearly labelled
1 culture tube containing 6 female Drosophila, either wild type (tube C) or mutant
  (tube B), clearly labelled
2 fresh culture tubes with medium
Labels
Etheriser, as described in the Experiment Guide, Section 11
White tile
Soft paintbrush
Hand lens or binocular microscope
Emergency etheriser, as explained in the Experiment Guide, Section 11

Comments

Drosophila experiments are time consuming and need careful preparation. Helpful
material, along with instructions, available from biological suppliers includes
prepared culture tubes, instant medium, virgin flies, prepared crosses, etherisers,
etc. Full details for keeping and using Drosophila will be found in Nuffield Biology,
Teacher's Guide V (1967 pages 21-5, or Revised Nuffield Biology, Teacher's Guide 4
(1977).

Specimen result

None

Think about it!

1   ... was the dominant characteristic.
2   This was dominant because all the F_1 flies had this characteristic.
3   The parent females had to be virgins to be sure that they had not mated with flies of the same type.
4   The F_1 females did not have to be virgins because they were to be mated with F_1 males anyway.
5   In the F_2 generation, the ratio of dominant to recessive flies was ....
6   The ratio of dominant to recessive over all the crosses was ....
7   The predicted ratio was 3:1. It could differ if there were too few offspring in the F_2 generation.
8   The results of the groups who used wild type males did not differ too much from those of the groups who used wild type females.

Suggested conclusion

Only wild type flies appeared in the F_1 generation, so wild type is dominant.
The F_2 generation yielded roughly 3 dominant to 1 recessive (the predicted result).
The larger the sample, the nearer it was to the predicted result. The sex of the flies had no effect.

12.1   Soil particles

Object

To show the existence of particles of different sizes in soils and to compare the proportions of these in different soils

Material per class

Clearly-labelled soil samples from different locations, eg pine wood, deciduous wood, garden, field, dunes, etc.(Large lumps should be broken up.)
Set(s) of soil sieves with the particle size each retains clearly marked on it. These are available with aperture sizes down to 0.038 mm
Balance(s)

Material per group/pupil

Tall, narrow container, eg coffee jar, gas jar with cover, or large boiling tube with stopper
Porcelain dish

Specimen result

None

Think about it!

Answers will vary depending on soils analysed.

Suggested conclusion

This could be a general conclusion stating the origin of the soils sampled and which types of soil they turned out to be.

12.2   Water and humus in soils

Object

To measure the water and humus content of a variety of soils

Material per class

Oven set at $105^{\circ}C$
Balance, as accurate as possible
Soil samples from the same locations as the last experiment, collected and sealed inside labelled polythene bags

Material per group/pupil

Pipeclay triangle
Crucible able to sit inside the
  pipeclay triangle

Tongs
Tripod stand
Bunsen burner

Specimen result

None

Think about it!

1   It is necessary to reheat and reweigh since the water or organic material may not all be driven off the first time.
2   If we did not, the results might be too low.
3   Clay should hold most water since it has the smallest particle sizes.
4   ....
5   Humus should increase the water content of the soil.
6   ....

Method A: Suggested conclusion

This could be an explanation of the results based upon the analysis carried out in the previous experiment.

Method B: Suggested conclusion

This could be an explanation of the results based upon the origins of the soils.

12.3 The air content of soils

Object

To measure the percentage air content of various soils

Material per group/pupil

2 tin cans with lids cleanly removed, one with holes punched in its base
Spade or trowel
Beaker or other suitable container large enough for the can to be sunk in it
  (Large plastic soft drinks bottles with the tapered part cut off are suitable.)
Marker
Measuring cylinder, 200 ml

Comments

It may be more convenient for the cans to be filled by the teacher beforehand.

Specimen result

None

Think about it!

1   If the can was not filled with soil, the extra air in it would give too high a result.
2   If the soil is compressed, air will be forced out of it and the result will be too low.
3   Stones will reduce the air content of the soil.
4   Filling the can with a trowel will disturb the soil and might affect the result.
5   Sandy soils should have most air in them and clay soils should have least.
6   ....

106

Suggested conclusion

The conclusion should relate the result to the type of soil, eg sandy or clay.

12.4   The pH of soil

Object

To estimate the pH of soils from various sources

Material per class

Distilled water
10 ml syringes
Barium sulphate
Spatulas for dispensing

BDH (or equivalent) soil indicator
    solution
Soils from various locations eg
    pinewood, field, garden, dunes,
    peat bog

Material per group/pupil

2 test tubes
Test tube rack
pH papers with colour card
2 stoppers

Glass rod
Colour chart for soil
    indicator

Specimen result

None

Think about it!

1   Tap water is not pure and this may affect pH.
2   In this experiment, using lime would change the pH.
3   Dirty apparatus may contain chemicals, etc, which might affect the
     pH.
4   ....

Suggested conclusion

The conclusion should relate the acidity or alkalinity of the soil to its source.

12.5   Soil properties

Object

To investigate and compare the water retention, drainage and capillarity of a clay soil, garden loam and sandy soil

Method A: Water retention

Material per class

Sets of 3 equal-sized, eg 10 cm diameter, glass filter funnels, supported by retort rings on stands. The spouts are plugged with glass wool and the funnels filled with equal volumes of dry sand, heavy clay or garden loam (enough room must be left for the addition of 50-100 ml of water). The soil samples should be dried in an oven at 90°C overnight and sieved through a fine mesh to remove the smallest particles. Large lumps should be broken up. Pottery clay can be used in place of clay soil.

Material per group/pupil

2 100 ml measuring cylinders

Method B: Drainage (permeability)

Material per class

Filter funnels filled with 3 types of soil as used in the last experiment. For this experiment they must have a line marked on them, all at exactly the same level.

Material per group/pupil

100 ml measuring cylinder
Beaker, 250 ml
Stopclock

Method C: Capillarity

Material per class

Lengths of glass tubing, about 20 cm long, of different bores from fine capillary
  tubing upward
Beaker of water, preferably coloured with eosin or potassium permanganate
3 glass tubes, 2 cm x 40 cm, closed at one end with muslin fixed with strong elastic
  bands. The tubes are filled up to 1 cm from the top with the same soils as used
  in the previous experiment. They are clamped vertically on retort stands.
Large beaker or trough of water

Specimen result (all methods)

None

Think about it!

1   Clay has the smallest particles, sand the largest.
2   The clay soil has the highest capillarity and sandy soil the lowest.
3   Clay soil has the highest water retention and sandy soil the lowest.
4   Clay soil has the slowest drainage and sandy soil the fastest.
5   Small spaces between soil particles show the highest capillarity.
6   High capillarity increases water retention and decreases drainage.
7   The garden loam has the best crumb structure.

Suggested conclusion

The small spaces between clay particles give clay soils high capillarity, while the
large spaces between sand particles give sandy soils low capillarity. This means
that clay soils have high water retention but drain slowly, while sandy soils have
low water retention and drain rapidly. Garden loams, because of their crumb
structure, have good water retention and drain well.

12.6   The effect of lime on soil

Object

To show that lime flocculates clay particles and also makes soils more alkaline

Material per class

Balance	Calcium hydroxide
Powdered potter's clay perhaps obtained	Spatulas for dispensing
from a friendly Art Department	

Material per group/pupil

2 large boiling tubes with stoppers
Test tube rack
pH papers with colour card

Specimen result

None

Think about it!

1   Clay particles are too small to settle out by themselves.
2   Lime causes the clay particles to stick together so that they sink.
3   Yes. We could see larger particles forming.
4   Lime could cause clay particles in a soil to stick together and so improve its drainage and crumb structure.
5   Lime is alkali and it will neutralise the acid in soil.

Suggested conclusion

The lime causes the tiny clay particles to stick together so that they sink. In a soil, this helps crumbs to form. Also, lime is alkaline and this will help neutralise an acid soil.

12.7   Collecting soil animals

Object

To collect, identify and estimate the numbers of organisms living on or in the soil

Material per class

Pooters
Pitfall traps
$\frac{1}{2}$ metre quadrat frames
Tullgren funnel
Baermann funnel
   (Versions of the above items can be
   bought from biological suppliers or
   made up as illustrated in Experiment
   Guide, Section 12.

10 litre batches of 2% formalin,
   dilute potassium permanganate or
   diluted washing up liquid
Forceps
Keys to identify soil animals

Comments

Only the minimum number of animals should be killed during this exercise. If a collection is to be made, only one or two specimens should be taken. The others should be identified, counted and released.

Specimen result

None

Suggested conclusion

A simple summary, in words, of the result with possibly a comparison of different habitats

12.8   Micro-organisms in the soil

Object

To show the presence of fungi and bacteria in soil

Material per class

Incubator set at 23-30°C
Fresh soil
Spatulas

Trough of disinfectant, eg 'Chloros' or 'Milton', freshly-made up. (Use rubber gloves when using these hypochlorite solutions.)

Material per group/pupil

1 sterile test tube, half-filled with sterile water and plugged with sterile
   cotton-wool
2 sterile test tubes, empty, plugged with sterile cotton-wool. (The test tubes are
   sterilised by autoclaving in a pressure cooker for 20 minutes.)
2 sterile Petri dishes containing nutrient agar (see page 8)
Marker to mark dishes
Test tube holder
Bunsen

Specimen result

None

Think about it!

1   We must raise the Petri dish lids only a little to prevent spores floating in
      from the air.

2   Boiling kills soil micro-organisms.
3   Any colonies growing in the 'boiled' dish could have come from the air.
4   Yes. Most growth is occurring round soil particles.
5   The micro-organisms would normally feed on the remains of dead plants and
        animals in the soil.
6   You must not touch the agar in case some of the colonies contain dangerous
        micro-organisms.

Suggested conclusion

Micro-organisms, both fungi and bacteria, are abundant in the soil. They are killed
by heat.

13.1   Food spoilage

Object

To show that exposure to the air causes food to go bad due to contamination by
microbes and that these microbes cause the food to disintegrate

Material per class

Clean fresh foods including bread, cheese, meat and fruit
Clean scalpels or knives

Material per group/pupil

5 small, 50 mm Petri dishes   Marker or labels
Water, preferably sterile, with dropper   Clear adhesive tape

Comments

For economy, each group need not do all 5 foods. Also, the teacher could set up
controls, under the most sterile conditions possible, for comparison with the
pupils' dishes. The time length can be increased if necessary.

Specimen result

None

Think about it!

1   The foods are being attacked by other organisms.
2   I can see ... different organisms.
3   Organisms are both fungi and bacteria.
4   The organisms are consuming the food. It is getting less and less.
5   The organisms have landed on the food from the air.

Suggested conclusion

Airborne organisms, both fungi and bacteria, land on food and consume it, making it go bad

13.2 Culturing micro-organisms

Object

To show the use of sterile techniques in the handling of micro-organisms and the appearance of agar cultures of bacteria and fungi

Material per class

Incubator set at 37°C. Some organisms such as Photobacterium phosphoreum have
  optimum incubation temperatures much lower than 37°C and their use reduces the
  risk of contamination with human pathogens.
Trough of disinfectant, eg 'Chloros' or 'Milton', freshly made up. (Wear gloves
  when using hypochlorite solutions such as these.)
Clear adhesive tape
Autoclavable disposal bags. (Autoclave all used Petri dishes and discard.)

Material per group/pupil

Bunsen burner
Clean white tile
Inoculating loop, made by bending 24 swg nichrome wire round a matchstick. The
  loop should be attached to a metal chuck holder since the end of this should also
  be flamed.
Inoculating wire - straight wire or wire with a right angle bend at the tip fitted
  in metal chuck
2 sterile Petri dishes containing nutrient agar (see page 8)
Nutrient broth culture of bacteria (see page 8 and Comments here)
Fungus culture on agar slope (see page 8 and Comments here)
Marker or self-adhesive label

Comments

The safety procedures outlined in Section 13 of the Experiment Guide must be strictly adhered to. If possible, protective clothing should be worn. Work areas should be away from draughts. Spillages on benches, etc, should be wiped with disinfectant. Spillages on equipment or the person should be wiped with an ampholytic surfactant disinfectant obtainable from biological suppliers. All cultures of micro-organisms, no matter how supposedly harmless, must be treated as potential pathogens. The organisms listed on page 114 are suitable especially those marked *.

Bacteria	Fungi
*Acetobacter aceti (-)	*Agaricus bisporus
*Agrobacterium tumefaciens (-)	Armellaria mellea
Bacillus subtilis (+)	Botrytis cinerea
*Chromatium sp	Botrytis fabae
Chromobacterium lividum (-)	Chaetomium globosum
Erwinia carotovora	Coprinus lagopus
Escherichia coli (-)	Fusarium oxysporum
*Lactobacillus casii (+)	Fusarium solani
*Lactobacillus bulgaricus (+)	Helminthosporium avenae
Micrococcus luteus (+)	Mucor hiemalis
*Photobacterium phosphoreum	Mucor mucedo
Pseudomonas fluorescens (-)	Myrothecium verucaria
Rhizobium leguminosarum (-)	Penicillium roqueforti
*Rhodopseudomonas palustris (-)	Phycomyces blakesleanus
Rhodospirillum rubrum (-)	Phytophthora infestans
Spirillum serpens	Physalospora obtusata
Staphylococcus albus (+)	Pythium debaryanum
Staphylococcus epidermidis (+)	Rhizopus sexualis
Streptococcus lactis (+)	Rhizopus stolonifer
Streptomyces griseus	Rhytisma acerinum
Streptomyces scabies	*Saccharomyces cereviseae
Vibrio natriegens	Saccharomyces ellipsoides
	Saprolegnia litoralis
(+): Gram positive,	Schizosaccharomyces pombe
(-): Gram negative	Schlerotina fructigena
	Sordaria fimicola

All cultures should be obtained from a reputable biological supplier. If there is likely to be any difficulty in following these precautions, the experiment is best done as a demonstration.

Specimen result

None

Think about it!

1   If they are not flamed, they may transfer unwanted (perhaps harmful) micro-organisms into the culture.
2   They should be flamed after inoculating the plates to kill any micro-organisms which remain on them.
3   The neck of the culture bottle should be flamed to prevent airborne micro-organisms getting in.
4   The Petri dishes are sealed to prevent any risk of infection.
5   The Petri dishes are stored upside down so that any condensation drips into the lid.
6   Bacteria form shiny-looking colonies, while fungi look fluffy.

Suggested conclusion

The bacteria form shiny looking colonies (growths), while the fungi look fluffy on agar.

13.3   Bacteria and antibiotics

Object

To test the sensitivity of bacteria to penicillin and streptomycin

Material per class

Incubator set at 37°C
Trough of disinfectant, eg 'Chloros' or
  'Milton', freshly-made up. (Wear rubber
  gloves when using these hypochlorite
  solutions.)

1 phial of penicillin discs
1 phial of streptomycin discs
Clear adhesive tape

Material per group/pupil

Bunsen burner
Clean white tile
Inoculating loop
Nutrient broth culture of bacteria
  (see page 8) - not E coli

Steril Petri dish containing
  nutrient agar (see page 8)
Forceps
Marker or self-adhesive label

Comments

The safety procedures outlined in Section 13 of the Experiment Guide must be strictly
adhered to.   Suitable bacteria are listed on pages 114 and 115 of this Guide.   The
cultures should be obtained from reputable biological suppliers.   However, all
cultures of bacteria, no matter how supposedly harmless, should be treated as
potential pathogens.   If there is likely to be any difficulty in following these
precautions, the experiment is best done as a demonstration.

Specimen result

Penicillin is not active against gram negative bacteria (see page 115), while
streptomycin is, streptomycin also shows some activity against some gram positive bacteria
such as Staph epidermidis and B. subtilis.

Think about it!

1   If the loop is not flamed, it may transfer unwanted (perhaps harmful) bacteria
     into the culture broth.
2   It should be flamed after use on the agar plate to kill any bacteria which
     remains on it.
3   The neck of the culture bottle should be flamed to prevent airborne bacteria
     contaminating it.
4   The lid of the Petri dish should be lifted only a little to prevent airborne
     bacteria getting in.
5   Antibiotics prevent the growth of bacteria since there were clear areas
     around some antiobiotic discs where the bacteria could not grow.

6   Antibiotics do not prevent the growth of all bacteria since some bacteria
     could grow round an antibiotic disc, while others could not.
7   Some antibiotics might be able to kill harmful bacteria that others could not.

Suggested conclusion

The antibiotic diffuses into the agar and prevents bacteria growing round it.
However, antibiotics are effective against different bacteria.

14.1   Investigating an ecosystem: collecting

Object

To illustrate various methods of collecting organisms from an ecosystem for
identification and counting

Material per class

Maps   Scoops, sieves, spades
Compasses   Mammal traps
Field notebooks/clipboards   Pen knives
Various nets   Polythene specimen bags with labels
Pooters   Vasculum
Pitfall traps   Plant press
Light traps

Comments

The equipment used and the results obtained will depend on the habitats investigated.
Most of the equipment illustrated can be obtained from biological suppliers.

Specimen result

None

Suggested conclusion

None

14.2   Investigating an ecosystem: plant distribution

Object

To study the distribution of plant species in various habitats by means of quadrats, line transects, belt transects and profiles

Material per class

Maps
Compasses
Field notebooks/clipboards
½ metre square quadrat frames
Stakes, mallet

Measuring tape, marked in cm, at
   least 10 m long
Large spirit level
Metre sticks
Keys to identify plants

Specimen result

None

Suggested conclusion

None

14.3   Monitoring water pollution

Object

To investigate the level of pollution in a stretch of water by looking at indicator animals, bacterial and oxygen content

Method A: Indicator animals

Material per group/pupil

Pie dish, preferably white enamel
Fine plankton net
Hand lens

Method B: Bacteria in water

Material per class

Incubator set at 37°C
Trough of disinfectant, eg 'Chloros' or
'Milton' freshly-made up. (Wear gloves
when using hypochlorite solutions such
as these.)

Clear adhesive tape
Autoclavabable disposal bags (Autoclave
all used Petri dishes and discard.)

Material per group/pupil

Bunsen burner
Clean white tile
Inoculating loop (see page 113)
Sterile McCartney bottle containing
tap water

Sterile McCartney bottle containing
polluted water
2 sterile Petri dishes containing
nutrient agar (see page 8)
Marker or self-adhesive labels

Method C: Oxygen in water

Material per class

'Polluted' water sample
Potassium hydroxide solution, 2 ml per group. Add 50 ml water to 100 g of potassium
hydroxide in a hard glass flask, mixing continuously. Keep the flask cool by
swirling under the tap.
Pyrogallol solution, 2 ml per group. Simultaneously, add 50 ml of boiled then
cooled tap water and about 25 ml liquid paraffin to 5 g of resublimed pyrogallol
in a beaker. Stir gently.

Material per group/pupil

2 gas burettes with Suba Seal stoppers
2 retort stands with clamps
Labels

2 1 ml syringes with needles
Trough of water

Specimen result

The 'polluted' water should be seen to be comparatively high in bacteria, low in
oxygen and to contain appropriate indicator animals.

Think about it!

1   Bacteria feed on organic waste and so multiply in number.
2   The bacteria use up all the oxygen in the water.
3   Fish cannot survive in water rich in bacteria since the presence of bacteria
     deprives the fish of oxygen.

4   Haemoglobin absorbs oxygen efficiently so blood worms can survive in water very low in oxygen.

Suggested conclusion

A statement of the degree of pollution in the water and the reasons for making it, eg high bacterial content, low oxygen, certain indicator animals

14.4   Air pollution: smoke on leaves

Object

To compare the accumulation of soot on pine needles from one to three years old and to compare the amounts of soot on the leaves of trees from different locations

Material per class

Branches of Scots pine with their place of origin clearly marked and having leaves up to 3 years old (see Experiment Guide, Section 14)

Material per group/pupil

Wooden clothes peg, spring-clip type   Strips of smooth paper about 2 cm wide
Matchstick, if necessary, with scalpel   Clear adhesive tape, scissors
   to trim it

Comments

A similar experiment can be done using the leaves of other evergreens such as holly. As before, the age of the leaves is determined by the position of the scars on the branch. A strip of clear adhesive tape is pressed on to the upper surface of the leaf to pick up soot, then the tape is stuck into the Student Record.

Specimen result

None

Think about it!

1   Older leaves were dirtier since soot has built up on them year after year.
2   ....
3   Soot on a leaf could cut down a leaf's light and clog its stomata, blocking its carbon dioxide supply.
4   Some leaves on a tree may be sheltered by others from soot or they may be on a different side of the tree from the source of pollution.

Suggested conclusion

The older a leaf becomes, the more dirt collects on it.   Some locations are dirtier than others.

14.5   Sulphur dioxide and plants

Object

To show the effect of sulphur dioxide on cress seedlings

Material per class

Sodium metabisulphate solution, fairly
   strong, freshly-made up. (Keep away from
   skin.   Do not inhale fumes or swallow.)

Tongs
Water with droppers
Cress seeds

Material per group/pupil

2 Petri dish bases, 30 or 50 mm
Blotting/filter paper
2 labels
2 watch glasses or Petri dish lids

Scissors
Cotton-wool
2 polythene bags
2 elastic bands

Specimen result

The seedlings exposed to sulphur dioxide should quickly appear affected.   After a day, they should be dead.

Think about it!

1   Seedlings B were not damaged.   We set them up as a control so that we could
      be sure that it was the sulphur dioxide which had harmed seedlings A and not
      something else.
2   Sulphur dioxide is harmful to plants.
3   Sulphur dioxide is produced by the burning of fossil fuels.
4   Sulphur dioxide is an invisible gas.

Suggested conclusion

Sulphur dioxide is an invisible gas, formed by the burning of fossil fuels.   It kills plants.